The SCIENCE of
FOOD

THE SCIENCE of

FOOD

*An exploration of what
we eat and how we cook*

Marty Jopson

Michael O'Mara Books Limited

First published in Great Britain in 2017
by Michael O'Mara Books Limited
9 Lion Yard
Tremadoc Road
London SW4 7NQ

A CIP catalogue record for this book is available from the British Library.

Papers used by Michael O'Mara Books Limited are natural, recyclable
products made from wood grown in sustainable forests. The
manufacturing processes conform to the environmental regulations of the
country of origin.

ISBN: 978-1-78243-838-0 in hardback print format
ISBN: 978-1-78243-863-2 in ebook format

1 2 3 4 5 6 7 8 9 10

Illustrations by Emma McGowan
Designed and typeset by K DESIGN, www.k-design.org.uk

Printed and bound by CPI Group (UK) Ltd, Croydon, CR0 4YY

www.mombooks.com

For my mother,
who taught me to cook

Contents

Introduction   9

The Essential Technology of the Kitchen   13

The Magic of Processed Food   58

Critical Kitchen Chemistry   100

Sharing Our Food With Bugs   133

The Future of Food   163

Acknowledgements   217

Index   219

Introduction

When I was a child I learnt to cook mostly from watching my mother in the kitchen. She was a cookery teacher and even when I was not actively helping prepare the ingredients or stir the sauces, I would sit on a stool, and watch and learn. I discovered how to handle food, how to use the paraphernalia of the kitchen and how to follow a recipe. At the same time, another interest and passion was beginning to grow, but it's much harder to nail down where exactly my love of science came from. I usually credit my grandfather, who sent me lots of *Reader's Digest* encyclopaedias, and my father, who seemed happy to take me on repeated trips to the Science and Natural History museums in London. At the time, I had no idea how much of a crossover there was between food and science, although my first foray into solo cooking demonstrated this link admirably.

This is one of those oft-repeated stories that has gained a level of notoriety in my family. It's wheeled out whenever it can be used to cause embarrassment and is guaranteed to result in much eye rolling. It also highlights how an understanding of science is what makes cooking possible. For reasons that are now lost, my mother had an errand to perform and I was to be left alone for a couple of hours in the house after school.

I told her I would be bored, so she said I should bake a cake. I am pretty sure she was joking and didn't really believe that I would. Once alone, I found a recipe book (a Delia Smith one) and picked out a method for Victoria sponge. Needless to say, I completely trashed the kitchen in the process of making the cake. As I recall, there was flour, egg and butter smeared all over the place, but I had helped make cakes before and confidently ploughed on regardless. I came across one major problem with the recipe: the baking temperature said 350° and our oven only went as high as 250°. I remember thinking this was odd but shrugging it off as a typo. I shoved the oven up as high as it would go and hoped the cake would cook. Some thirty minutes later, I was frustrated when I pulled the cake out to discover the outsides charred black. I must have been about ten years old, so I can't really be blamed for having no concept of imperial or SI units. Apparently, back in 1978, cookbooks only gave temperatures in Fahrenheit, and our oven only listed Celsius. Undaunted, I scraped the burnt bits off, made up some icing and drizzled it over. I even managed to return the kitchen to what I considered to be a pristine state. My mother maintains the cake was delicious, but I recall it was barely edible.

The point of this anecdote is that to make edible food it was not enough to follow a recipe. Despite my best efforts the cake was still a bit of a disaster. I had no idea that temperature could be measured on two scales or how to convert from one to the other. If I had known a bit more about the science of temperature, maybe I would have spotted my error and could have applied that knowledge to make a better cake. Cooking is about the appliance of science, whether you are aware of it or not. It is, of course, possible with no understanding of what is going on to learn to cook delicious meals, but you will be cooking by rote. If you step away from the things you know, or when things start to go wrong, you have no way to navigate back to a successful result if you don't understand the processes involved.

There is also a huge and wonderful world of science behind the food we don't prepare for ourselves. The processed food we buy from the supermarket is full of some of the most ingenious science I have ever come across. I was lucky enough to spend three years working on a television series about processed food, researching and then building machines to replicate industrial food processes. I made a processed bread machine in a dustbin, a salmon smokehouse in a cheap flat-packed cupboard and a bathtub became a milk pasteurization device. Like my early cake baking, not all of these devices were a success; my attempt to use a 1960s mangle to make shredded wheat cereal was a complete flop, and I can no longer look at a mangle without recalling the stress caused by trying to get the cursed machine to work. However, my favourite was the machine that could crack fifty eggs in one go and would then separate the yolks, all in about fifteen seconds.

I have tried in this book to capture a little bit of the science that plays such a huge role in the production of food that you find on the shelves of your supermarket and in the meals that you prepare in your own kitchen. Taken together, I hope I've cooked up something that gives more than just a taste of the science of food.

The Essential Technology of the Kitchen

The thin edge of the wedge

If, like me, you are a fan of gadgets you have probably accumulated a number of peculiar devices in your kitchen drawers and cupboards. I have one drawer in particular that resists being opened due to all the kitchen technology crammed into it. Some choice items contained in this recalcitrant drawer include: the milk-foaming whizzy thing that was only used twice, the wine bottle vacuum pump for half-finished bottles, and the twice-as-fast mandolin that cuts both ways and slices your fingers twice as efficiently. A quick survey of all my gadgets reveals that they generally fall into one of two types: things for preparing food and machines for cooking food.

The food-cooking machines tend to be bigger and are geared to different methods of cooking that are, on the whole, only possible with these machines. So, the slow cooker contains a thermostat without which such prolonged cookery would be impossible, and the bread machine turns the production of a loaf into a ninety-second prep-and-ignore activity. The hot-air popcorn thing is mostly retained for the amusement value of seeing the kids trying to capture puffed corn as it flies violently out of the open mouth of the machine and ricochets around the kitchen.

However, when it comes to the food-preparation devices – the mandolins, peelers, crushers, dicers and chip-makers – I have a sneaking suspicion that every single one of these is redundant. With a bit of practice, all of these gadgets can be replaced by a really good knife. Surely, the knife is the ultimate in kitchen gadgets; an irreplaceable tool for the cook, and the most versatile.

I have a modest collection of kitchen knives. My current favourite knife is a wonderful Japanese-style Santoku with a cherry-wood handle. It holds a beautiful edge, cuts through anything like butter and suits my cutting style. But why does a knife cut in the first place? And can an understanding of this influence knife usage in the kitchen?

If you consider how a knife is used, it has two basic modes of operation. First, there is the classic chop, which entails a vertically straight down movement of the blade through the food. Secondly, you have the slice where the blade of the knife is drawn across and down at the same time as it cuts. While the chopping action is ideal for some things, like cheese and carrots, for others the slice is much easier than the chop. How can it be that the same knife cuts some items better when slicing than chopping?

As an extreme example, consider the painful yet all too common paper cut. A sheet of paper is quite useless when it comes to chopping your finger, but if you run your finger along the length it appears that it can readily slice into flesh.

The answer to this conundrum is all to do with shear and has been studied in some detail in the laboratory. The basic concept behind cutting anything is that you are producing a fracture and then forcing that fracture to propagate through the material being cut. Creating that initial fracture is the hardest part and, once made, the split can be pushed forwards through the material much more easily. All material, be it an apple, a chicken breast, a block of cheese or a lump of wood, has an inherent resistance to fractures. The molecules that make up the object are hanging on to each other and resisting the intrusion of the knife. Until, that is, the stress applied by the knife between the molecules gets to be greater than the force holding the molecules together. At that point, they snap apart and we have created a fracture. So, key to cutting is creating the initial fracture by increasing the stress between molecules.

This was wonderfully tested by a bunch of researchers at Harvard University back in 2012. They carefully measured the forces and stresses applied to a series of small blocks of agar jelly as they were cut with a tautly stretched, very thin wire. The force needed to create the critical level of stress in the jelly block when they tried chopping was more than twice needed for the slicing action. On a microscopic level, as the sharp edge slides across the object to be cut, it catches on it, effectively sticks to it and creates friction. This friction pulls the surface sideways, creating a shearing force as well as the downward force. Combined, these are enough to initiate a fracture and the cut can then propagate.

This is why paper, which cannot chop skin because the paper is all floppy, can still slice. If you slide your finger along the edge of a sheet of paper, the paper itself is pulled taught and acts as a knife blade. The very edge of the paper is rough and creates lots of friction and enough stress in your skin to start a fracture. Once begun the paper can then elongate this fracture, creating a cut. Interestingly, the reason paper cuts are so painful is due to the relative roughness of the edge of a sheet of paper when compared to a sharp knife. The paper edge creates a ragged tear in the skin, causing more tissue damage and more pain than a sharpened metal edge.

This helps us understand why the recommended way to use a knife is with a gentle forward motion along with the downward push. This way you are creating a slicing motion rather than a chop and the effort needed is much reduced. Why then do we still chop a carrot and a block of cheese? In the case of the cheese, the material is sufficiently soft that the blade easily pushes into the block and starts the fracture going. Carrots on the other hand are so brittle and their cells large enough that the blade of the knife can get the fracture started with little effort.

Once you have initiated the fracture, you then want a thin wedge of a blade to split that fracture and propagate it through the material, creating a cut. So, the knife actually needs to do two jobs. Conveniently for us, the best way to do this is to have a devilishly sharp edge on your blade. When looked at under a microscope, a sharp blade is not as smooth as it may seem. Instead, it consists of a series of ridges and furrows running up to the blade edge, creating what is to all intents and purposes a microscopically serrated edge. As this edge slides across food it catches, creates the needed friction to produce the shearing force that increases the stress that initiates a fracture. A blunted blade, on the other hand, has a rounded and smooth edge that slides, without catching, across food and does not start a cut so easily. Consequently, since you have no shearing force to help, you have to rely solely on the chopping action and need to apply much more force. Which is why blunt knives are more dangerous than sharp ones. All that extra force means you are more likely to slip and that's how accidents happen.

Given the complexity of the task a blade is performing, with all the shearing forces and friction needed, it should come as no surprise that the manufacture of a knife is also a smidgeon complicated. To create a blade that can hold a sharp edge you want to use really hard steel. But on top of this you want the edge of the blade to be resistant to being worn down and for that you need a tough steel. Crucially for a knife, and any material scientists, hardness and toughness are not the same thing. Hardness is the ability of a material to resist being scratched or deformed by compression. Toughness is a measure of how well a material can absorb energy and deform without breaking, or, to put it another way, how well it copes with being bent. In a knife blade, you want your steel to be hard so that the edge stays there; in addition it should be tough so it doesn't get worn down and the blade won't snap the first time you flex it a bit. This is the tricky bit, as an increase in hardness usually reduces the toughness, and tough steel tends to be not so hard. Clearly, it's a balancing act so knife manufacturers add carbon to the iron metal to create hard steel, tungsten and cobalt for toughness, and a spot of chromium to make it stainless and prevent rusting while they're at it.

The final part of knife science I need to mention is the angle of your wedge. A standard Western- or Germanic-style knife blade will be sharpened so that the angle between the two sides of the blade is about 35 degrees. But the Japanese Santoku-style blades are much finer with a total angle of only about 25 degrees. The fineness of the blade makes a big difference to the edge you can put on a knife. Finer blades give a sharper edge and will thus cut easier and with less effort. So, why not make all blades as fine as possible? Well, this comes down to practicality and what the knife is being used for. Santoku blades, while being sharper, are more prone to being dented and bent in use and in storage. If you are using a 25-degree-angle blade and accidentally come across something hard in what you are cutting, like a bone for example, there is a good chance you will damage the blade. Similarly, if you want to keep your Santoku blade in good condition, don't slip it in the kitchen drawer crammed with gadgets. Broader, 35-degree blades don't suffer these problems, but will never take an edge quite like a Santoku.

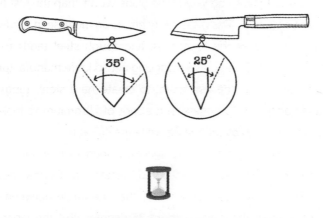

Chop, chop, chop

What use is a wonderfully sharp and sleek kitchen knife without a chopping board? The board is the less glamorous but equally important part of this ubiquitous duo, yet even here there is hidden science for the unsuspecting.

The key issue when it comes to the design of a chopping board is hardness of the board material: its ability to resist being deformed by compression, or specifically its resistance to being cut. Too hard, and it will blunt your knives. Conversely, if it's too soft the board would fall apart.

To get a sense of how hard is too hard and how soft is too soft, we need to quantify hardness. There are several ways to

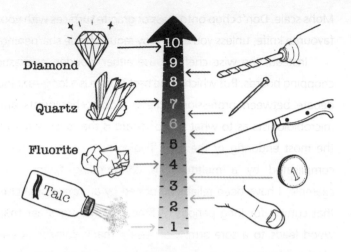

Diamond

Quartz

Fluorite

Talc

do this, but the simplest is to use the Mohs scale of hardness, created in 1812 by a German chap called Friedrich Mohs. The Mohs scale goes from 1 to 10 and was really created to quantify the hardness of minerals. In particular, any mineral with a higher rating on the scale was able to scratch those lower down. Diamonds are at the top of the scale with a 10 and they can scratch anything below them, such as quartz at 7 for example. Similarly, quartz will scratch gypsum since this is only 2 on the Mohs scale.

The steel used to make the blades of knives is in the order of 5 or 6 on the Mohs hardness scale. This means you should never use a chopping board harder than that. Note that both glass and granite kitchen countertops, which are primarily made of quartz, have a hardness of 6 and 7 respectively on the

Mohs scale. Don't chop onto glass or granite surfaces with your favourite knife, unless you also enjoy regular blade sharpening.

Instead, the wise chef will use either wooden or plastic chopping boards. But which is the best? There is a long-running debate between professional chefs, food technologists and microbiologists as to what sort of board is the most practical, the most enduring or the most hygienic. It quickly becomes complicated by a multitude of confounding factors. For example, I have been reliably informed by a professional chef that cutting for long periods of time on anything other than wood leads to a sore arm. Conversely, many domestic users of chopping boards prefer plastic because they don't have dedicated cleaning staff, and the board can be chucked in the dishwasher. Then again, some people claim that the natural phenolic compounds in wooden boards actively kill off bacteria lingering on the surface. Which leads me nicely onto one of the most crucial aspects of chopping board science: hygiene.

Since you are invariably placing raw food on the board, the potential for bacteria to remain behind and contaminate the next thing on the board is a real risk. Clearly the most obvious thing to do is to follow the lead of all commercial kitchens, which use a separate chopping board for items such as raw meat that contain the highest potential for harbouring nasty bacteria, including salmonella.

In an effort to go beyond anecdotal arguments, several scientific studies have been carried out, including one I was involved in for a TV series that I presented. In a rare example of properly controlled TV science, the tests were carried out by an accredited laboratory of UK government scientists based in Glasgow. We started out with a big pile of new and used chopping boards, some made of wood and some from plastic. First up, to give us a uniform hygienic baseline, the boards were all identically sterilized. We then contaminated sections of each board with solutions containing a known number of bacteria. The boards were air-dried and then sampled over the course of twenty-four hours. The number of bacteria in each sample was then worked out by laboriously smearing a bit of each sample on a Petri dish, leaving it to mature and then manually counting the bacterial colonies that had grown up.

The aim of this part of the test was to simulate putting something such as raw chicken onto the board, then failing to clean it properly – perhaps giving it a perfunctory wipe – and using the board again some time later. We were deliberately seeing if we could test the idea that wooden boards were in some way anti-bacterial. Would the wood kill off more bacteria than the plastic? To the disappointment of the director on the day, the answer was no; in fact, it made no difference at all what the board was made of or how old the board was. Uncleaned boards retained a disturbingly large number of bacteria.

So, what about if you actually do what you are supposed to do and clean your board after you have used it? Once more, we set to with our chopping boards, but this time after they had been inoculated with bacteria, we gave them a thorough scrubbing with hot soapy water. The boards were tested for bacteria one last time and once again there was a resounding failure to find any significant difference.

From a television perspective, this was a bit of a disaster. We had set up the huge scientific test, explained all the complicated procedures and our result was a thoroughly disappointing lack of difference. Which was a little surprising and flew in face of several of the previous studies made on chopping boards.

From a scientific perspective, what this indicates is that if there is a difference between wood and plastic chopping boards, it is marginal and probably more influenced by the exact cleaning protocol used rather than the chopping boards themselves. If this is the case, the implication for the home cook or even professional chef is that you should use whatever board you fancy. If you want something that can go in the dishwasher, go plastic, but if you prefer the feel or the aesthetics of wood, go with that instead.

However, all studies agree on one thing: if the surface of your board gets really hacked up with deep grooves, then it becomes a serious health hazard as no matter how much you scrub, it will never come clean and bacteria will fester in the grooves. And another thing, if your wooden board starts splitting then you won't just be harbouring bacteria, but chunks of food as well. I would also advise against using bamboo for chopping boards. While it may look and feel like wood, bamboo is actually a grass, and grass is particularly good at producing little shards of silica called phytoliths in its stems. And silica is harder than steel, so a bamboo chopping board will blunt your knives just like glass. As you can see, the business of choosing a chopping board is a complicated task.

What about ceramic?

Now that I've introduced some of the issues facing chopping-board choice, it's worth mentioning the newest technology to hit the kitchen cutting scene. It's now possible to make a knife blade from a ceramic. While this conjures images of a porcelain knife blade, which we can all agree would be rubbish, the material in question is much more high-tech. The blade of a ceramic knife is an otherworldly substance: very hard and light, almost translucent and with a razor-sharp edge. The blade is made from zirconium dioxide, or zirconia; the same stuff used to make the cubic zirconia gemstones found in jewellery on late-night shopping channels.

At its simplest, to make a ceramic knife you take powdered zirconia, press it into a knife shape and then heat it up to fuse the powder together. Which makes it all sound like a project for a science fair. The reality is that you need pressures in the order of 900 atmospheres, or one tonne for every square centimetre (14,200 psi), and a temperature of 1,400°C (2,550°F). At this pressure and temperature the fine zirconia powder fuses together to form a solid. The process is properly known as sintering and is the same process that takes snow and turns it into an icy glacier. Once your blade is sintered and sharpened it's ready to be put through its paces. A ceramic blade has a big advantage over steel blades as the sintered zirconia has a hardness of 8.5 on the Mohs scale, which makes it harder than steel or glass or pretty much anything naturally occurring except for diamond. This means that it holds an edge far longer than a steel blade – ten times as long according to one manufacturer.

So, clearly we should all be using ceramic blades and chuck out all the rubbish steel. Well, not so fast there. The very hardness that makes the blade so enduring is a problem. To sharpen any blade, you need to use something harder than the blade, and that means a diamond-dust coated tool for a ceramic blade. It's also much trickier to sharpen a ceramic blade so manufacturers either recommend you send the blade back to them for sharpening, or just throw it away and treat it as a consumable item. That extreme hardness will also cause problems with your chopping board. A ceramic blade will slice up any surface you cut on, leaving marks on glass and even granite worktops.

On top of that, there is another major issue with ceramic blades. As with so many things in life, the extraordinary hardness of zirconia comes with a compromise. As the hardness goes up so the toughness goes down, and this is where we once again come up against those cheeky material scientists and their very specific use of common words. As we have seen, toughness is the ability of a substance to absorb energy and not fracture. Steel is pretty tough and if you try to bend a steel knife it will flex and return to its original shape. Put more energy in and it will eventually give way, bending and changing shape. Ceramics, zirconia included, are not very tough and if you try to flex or bend a thin sheet of ceramic it will crack. If your lovely ceramic blade hits a bone or an unexpected hard bit and you twist the knife, a little bit of your super sharp and hard edge will snap off. Worse still, drop the knife on the floor or sling it carelessly into a drawer with lots of other utensils and there is a good chance it will snap in two.

Ceramic blades are currently viewed with some suspicion by cooking professionals. They are undeniably sharp and remain so, even without regular honing, but their fragility makes them less useful as a multi-purpose tool. You will notice I say currently, as there is a constant stream of new ceramic formulations being developed that provide new material properties. But it is unlikely that a ceramic blade will ever beat a steel one on toughness and hardness. If you have a ceramic blade, save it for those delicate tasks at which it will excel. And try not to drop it.

Why temperature matters

Equally complicated and confusing can be the task of working out how to supply the heat needed to cook the food you have so carefully chopped and sliced and peeled. There is a huge panoply of ways to add heat to food, from the simplest barbecue, through grills, fryers, slow cookers, ovens, microwaves, induction hobs and the newest, hippest, most scientific way: the sous vide cooker.

All of these heating devices and machines are trying to do one thing: change the temperature of the food you are cooking. Now, I appreciate that this may be the most stupidly obvious statement anywhere in this book, but bear with me for a moment. The act of cooking anything is about changing the temperature of the food so that one of a variety of biochemical reactions can take place. Which biochemical reaction you are trying to make happen depends on exactly what you are cooking and what you are trying to achieve in the way of flavour and texture. There are really only three categories of food that you may be playing about with: sugar, starch and protein. The first two of these I'll get to in other chapters of this book, but I wanted to discuss protein now as this is where the most interesting new technological developments are being made. I have not included fats in the above list because, while the melting temperature is important, you are rarely trying to chemically change the fats through the process of cooking.

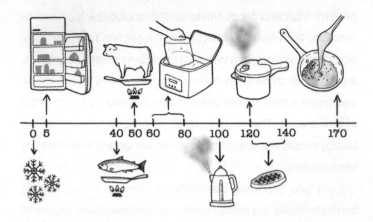

So, consider a lump of protein that you are trying to cook. This may be a steak, a piece of fish or even an egg. The end result is that you are trying to take the protein molecules from their normal or native state to a heat-changed form known as denatured. To understand this, we need to remind ourselves of some basic protein science. All proteins are made up of chains of a family of chemicals called amino acids. Key to the whole family is the presence of at least one nitrogen atom per amino acid and there are generally only twenty different types of amino acids in proteins. The order of amino acids in the protein chain is what makes proteins different from one another. So, the protein called ovalbumin, which makes up the bulk of an egg white, has a chain of 385 amino acids always in a particular order. On the other hand, 55 per cent of all muscle fibres in something like a steak are made up of a protein called myosin, which has some 2,000 amino acids in their own unique pattern. It is the order of amino acids that gives each protein its function, but also determines how the protein folds itself up. Since many amino acids will create bonds to other amino acids, any chain of these chemicals will spontaneously fold itself up and create a blob, whose shape is also determined by the order of amino acids. The native state of any protein is this folded-up blobby shape. But that's not what we eat once the protein has been cooked.

As you gradually warm the protein, the heat energy begins to shake the blobby molecules and eventually breaks all the bonds between amino acids. This is the point at which the protein denatures. It uncoils itself from its balled-up shape and turns into a freely wiggling bit of spaghetti shape. What then invariably happens is all those wiggly spaghetti molecules stick to each other. Once it has denatured, the overall texture and colour of the lump of protein changes and we would consider it cooked and more easily digested by us. This is the crucial bit for cooks: the temperature at which a protein goes from native to denatured depends on the bonds inside it and is thus unique to the type of protein. Which is why you need less heat to cook fish than you do to cook meat. Myosin from a salmon is a little bit different to myosin protein from a cow. They both do the same job within the animal, but subtle amino-acid differences mean that salmon myosin starts to denature at 40°C (104°F) while beef steak myosin starts at 50°C (122°F).

Heating things up

Understanding the physics of how changes in temperature affect our food is one thing, but what about the science of how you go about doing this? The direct application of heat requires some sort of saucepan or frying pan in which to heat your food. It might seem that this would be a relatively straightforward process, but if you venture into a shop looking to buy such an item you are confronted with a bewildering array of choices. Once you remove cosmetic details, the real issue is what do you want your pan to be made from? You can choose from steel, aluminium, copper, cast iron or even layered combinations of these materials. As with knives (see page 13), the choice comes down to the physical properties of each material in question. In this case one of the key properties is the ability of different metals to transfer heat, and the scientific term for this is thermal conductivity.

Not all metals conduct heat as well as others. Copper is one of the best but it is somewhat surprising that stainless steel is a really poor conductor of heat. Heat conductivity is really important in a pan as the source of heat is often not even across the bottom of the pan. Gas hobs in particular apply a ring of heat with an unheated spot in the middle. If you make a pan out of a highly heat conductive material, like copper, the heat quickly disperses across the entire base giving a uniform heating surface. On the other hand, a pan made from stainless steel, especially if it's thin steel, won't give an even heat and, in extreme circumstances, will have hot spots that burn your food. It would seem that copper is the best material to make pans from, but pure copper is very rarely used for a few reasons: it's expensive, it tarnishes easily and in acid conditions it dissolves into food to toxic levels, so it can't be used with things like tomatoes or lemons. Pure copper has only one real niche application and that is in bowls for beating eggs (*see* page 49).

The next best material we have in terms of ability to conduct heat is aluminium and you do find many pans made from this, but it is not a perfect solution. While it does make for very lightweight pans, once again it will react with acid foods. In this case, it's not a toxicity issue but the dissolved aluminium can turn your food an unappetizing grey colour. Aluminium does have one special advantage that makes it a popular choice with chefs, and that is its ability to hold onto more heat than an equivalent weight of copper. This property is known as specific heat capacity and is measured as the amount of energy used to heat one kilogram of material by 1 °C. Aluminium has nearly three times the specific heat capacity of copper, which means it heats up more slowly, but it also cools down more slowly. This property makes it ideal for frying pans. If you are trying to rapidly cook a piece of meat, an aluminium pan will cool down more slowly and you can more effectively sear the meat, creating delicious Maillard reaction products (*see* page 100).

Carrying on down the line of thermal conductivity, and just a bit better than stainless steel, is cast iron. But as anyone who has owned a cast-iron pan knows, the issue here is rust. If your pan is not properly dried after washing it is going to rust, which, if not cleaned off, will ruin the food you next cook in it.

Finally, we get to the worst thermal conductor, stainless steel, which is somewhat paradoxically also the most used material for saucepans and frying pans. In the end, it is the convenience of stainless steel that trumps all other materials. It does not tarnish, needs no special treatment and is much tougher and thus less likely to scratch or dent in use. It is also the only material commonly used for pans that is magnetic, which is a big consideration as modern induction hobs only work with magnetic materials. An aluminium pan is useless on an induction hob.

Fortunately, material science can come to the rescue of the cook wanting to have both the thermal properties of aluminium or copper and the durability of steel. Many pans are now made with more than one type of metal. The simplest way this is done is with what are known as copper-clad pans. Manufacturers take a sheet of steel, a sheet of copper and then another sheet of either steel or occasionally aluminium. This sandwich of metal sheets, with copper in the middle, is run through a very hot system of rollers that squeezes and fuses the sheets together into one. Pans made from these sheets have steel on the outside not only for durability but also so that they work on inductions hobs. The copper sandwich filling helps spread the heat around. On the inner surface is either another layer of steel or, in frying pans, aluminium for its specific heat capacity.

The alternative to clad copper are copper-core pans, which are usually more expensive as the process is more difficult to do and uses more metal. These have a base made from a flat disc of copper encased in aluminium, which is then encased in steel. The resulting thick base disk is fixed to the bottom of a steel pan, which now has the advantages of copper's high heat conductivity, aluminium's high ability to hold onto heat and steel's durability.

You now have a frying pan or a saucepan with the perfect combination of materials to give a uniform and long-lasting heat, but when you cook with it food sticks to the surface. Chemically, what happens here is that proteins and sometimes sugars are reacting with the surface molecules of metal in the pan. This happens with copper, aluminium and steel pans, and the simplest way to stop this is to stir the food, keeping it moving and not giving the chemical bonds time to form. Failing that, coating the metal with something less reactive will also prevent sticking, and the most common non-stick coating used is Teflon.

Invented by accident in 1938 by an American chemist called Roy Plunkett, Teflon or polytetrafluoroethylene (abbreviated to PTFE) is a long carbon molecule made unreactive by the addition of lots of fluorine atoms. The problem is how you stick something very non-sticky like PTFE to the surface of a frying pan. Chemical methods have been used to do this, but the chemicals involved are unpleasant and toxic. Instead, these days the pan to be coated is first sandblasted to create an incredibly rough metal surface. When liquid PTFE is then applied, it flows into all the rough nooks and crannies made by the sandblasting. As the now smooth surface hardens, it is physically bonded to the underlying material. The PTFE grabs onto and holds all the little bumps and lumps of metal. Once the coating is securely stuck down, the secret to the success of PTFE is the wall of fluorine atoms at the surface. Fluorine bonds incredibly strongly to the carbon in PTFE and, once bonded, cannot bond to anything else. Consequently, the food in the frying pan has nothing to attach itself to and won't stick.

There is a lower-tech version of Teflon that is the saving grace for cast-iron pans. To make a cast-iron pan non-stick, or to season it, you need to first coat it in a thin layer of oil and then bake it in a very hot oven (260 °C or 500 °F) for about an hour. This intense heat breaks down the oil into small two- or three-carbon atom units. As you then cool the pan, these units link together forming hugely long carbon chain molecules. These long carbon chains act like PTFE, coating the underlying metal and preventing food from making chemical bonds with it. It doesn't have the unreactive fluorine coating but it has the advantage of being scratch resistant and easy to reapply.

When you come to choose a frying pan, be it a new one from a store or maybe just one from the selection in your cupboard, take a moment to consider the science. Critical to what makes one pan better than the other is the material science of the component metals, and also the chemistry of the surface layer. Get this correct and you have a perfect cooking surface every time.

Under vacuum

Of all the ways you can cook protein, the sous vide method relies more than any on the digital precision of a temperature probe. First, you place your food in a plastic bag, then suck all the air from the bag with a vacuum, seal the bag and finally place it in a digitally temperature-controlled water bath. The name sous vide comes from the French and means 'under vacuum'.

You may be thinking at this point that this is just a fancy and overly complicated way of poaching something. You may have a point, but there are two things that set sous vide cooking apart from simple poaching. Firstly, the food is sealed in a bag containing no air. Any flavour or moisture from the food stays in the food and is not wafted away in the poaching water. The same is true of added spices and herbs you put in the bag before sealing it. Furthermore, the absence of any air in the bag stops spoilage from oxidation and if the cooking temperature is high enough the contents are effectively sterilized by the process, so the food can be stored in the bag.

The second big benefit of sous vide cooking is that the temperature of the water bath is never boiling; in fact, it is rarely used above 80°C (176°F) and more often it is set to somewhere around 60°C (140°F). It is the temperature of the water bath that is critical to the mouth-watering results of sous vide cookery. The water bath temperature is usually controlled to within a fraction of a degree. It's not a complicated machine as it consists only of a heating element controlled by an ever-so-handy digital thermometer probe. You set the exact temperature you want: the probe monitors it and turns the heater on and off accordingly.

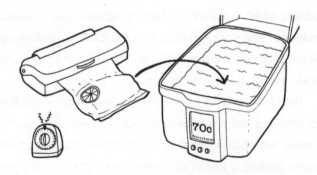

So, let's say you are cooking a piece of fillet steak. You set your sous vide water bath to 57°C (135°F), pop the steak in a bag, vacuum seal it and bung it in the water. Now, very slowly, taking about an hour, the temperature of the meat comes up to the temperature of the water bath, 57°C

(135 °F). At this temperature, most but not all of the different types of protein molecules in steak will denature. The myosin that makes up the bulk of the steak will be denatured, making the meat tender and not tough. Another protein called myoglobin, which gives the meat its red colour, will be just starting to denature so the meat won't be blood-red but pinkish. However, the actin protein will still be in its native form, which is good as when this denatures the meat turns tough and tastes less juicy. The whole block of meat, from the outside to the very centre, will be at exactly 57 °C (135 °F) and consequently a perfectly cooked medium-rare. If you want your steak rare, the temperature you need is 49 °C (120 °F), well below the temperature that myoglobin denatures. For medium, it's 60 °C (140 °F) and the myoglobin has completely denatured. And if you want to ruin the steak, at least in my opinion, a temperature of 74 °C (165 °F) will denature all the protein including the actin and give you well done.

This is the genius of the sous vide method of cooking and why it is the most scientific way to cook. By knowing at what temperature the different proteins in your food become cooked, you can achieve precise, reproducible levels of cooking. To take another example: the humble egg that so many of us struggle to cook the way we like it. Part of the problem is that we all like our eggs cooked differently. Are you a runny-yolk person and, if so, are you willing to put up with a bit of unset egg white? Or is the thought of floppy egg white so vile you take the egg the whole way and want the yolk just set, but not overset when it turns crumbly. The other issue with cooking eggs is that not all eggs are equal: they come in different sizes, varying freshness and the temperature before you start cooking makes a difference. When you plop an egg into boiling water, the outside is immediately at 100 °C (212 °F), which is a temperature that will denature all the proteins in the egg, setting the white and the yolk and even beginning to release some of the sulphur compounds in the proteins. Clearly the key to cooking the perfect egg in a pan of boiling water is timing it just right, taking into account size, freshness and starting temperature. Not so if you use a sous vide cooker.

Firstly, it should be noted that eggs are perfect for sous vide as they already come in a handy sealed package and need no vacuum treatment. The albumin or white portion of the egg is made up of a number of proteins, most of which cook or denature between 61 and 65 °C (142 and 149 °F). The yolk protein denatures and turns solid between 65 and 70 °C (149 and 158 °F). Based on this information you can now make the perfect sous vide cooked eggs. For just-cooked white and completely runny yolk set you water bath to 63 °C (145 °F). If you prefer a firmer white and a mostly runny yolk, it is 66 °C (151 °F), and for a just-set yolk turn it up to 70 °C (158 °F).

The beauty of sous vide cooking is that the temperature at which a protein from a particular animal denatures or cooks is always exactly the same. Knowing this, you can take the guesswork out of cooking the food the way you want it, and get reproducible results every time. So, why don't we all have sous vide cookers? Well, you actually need two pieces of kit, the water bath and the vacuum-sealing system, both of which tend to be bulky and expensive. On top of that, sous vide cooking gives a different type of result. A steak cooked sous vide may be perfectly medium-rare, but will have no delicious browned surface. The yummy outside of a steak is caused by the Maillard reaction that happens at 154 °C (309 °F) (*see* page 100). Finally, it takes a lot longer when you are cooking at such low temperatures for the heat to fully penetrate your food. While sous vide is in theory a brilliant way to cook and has its place, I can't see my family waiting an hour in the morning for a perfectly runny egg.

Under pressure

So if the sous vide method is the most scientific way to cook food really slowly, then how does a science geek go about cooking food fast? You may be expecting me to start on with the microwave oven at this point, but for me the pressure cooker wins the geek credentials. The story of its discovery is also rather intriguing.

Towards the end of the seventeenth century, a Frenchman called Denis Papin was working as an assistant to the Curator of Experiments at the Royal Society in London, the oldest scientific society still in existence. The curator was a somewhat irascible man called Robert Hooke who cast a huge shadow across the scientific landscape. Papin presumably had some leeway with his work as in 1679 he demonstrated to the assembled Royal Society dignitaries his 'New Digester or Engine for Softening Bones', or the pressure cooker to you and me. The device consisted of a metal pot, without a handle and a lid that could be screwed down to create an airtight seal. Crucially, he had also invented the safety valve that used a lever and a weight to prevent steam escaping through a hole in the pot's lid until the correct pressure was achieved.

As part of his demonstration he placed within his pot all manner of cheap cuts of meat and a little water. In no time at all he had produced a delicious, tender and succulent stew to the delight of the assembled scientists. In 1681, he published his culinary experiments and invention in a small pamphlet in which he explained how the pressure cooker could, among other things, be used to feed the poor with nutritious gravy made from the cheap meat nobody else wanted, as well as rabbit. He seemed to spend a lot of time cooking rabbit in his New Digester. Sadly, the Royal Society didn't really pay much attention to Papin's invention and it seems to have been viewed more as an academic curiosity.

At some point in the next 200 years, the pressure cooker moved from the hands of academics to the everyday cook. What history does not record is exactly how or when this took place. We know that in 1864, Georg Gutbrod of Stuttgart in Germany had a secret process to make cast-iron pressure cookers, coated in tin. His cookers were deemed to be superior to others available at the time, which clearly implies that the devices had already been in general use for quite a while. What marks these early devices from the modern variety is that they all look like pieces of industrial kit; they were incredibly heavy, had enormously thick walls and the lid was fastened with great big screws and clamps. Which probably goes some way to explain why they never became particularly popular. Then in 1938 an American chap called Alfred Vischer revealed a new pressure-cooker design that looked and handled much like a regular saucepan. Since then the design of the pressure cooker has remained essentially the same.

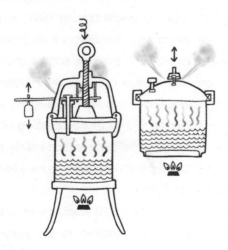

Papin pressure cooker (left) with screw-down lid and lever safety valve, and the modern pressure cooker (right).

The clever bit about a pressure cooker is that it makes the water inside the cooker boil at a temperature greater than 100°C (212°F). When a pressure cooker is up to steam, the water inside is boiling at around 120°C (248°F). Which might be surprising, as everyone knows that water boils at 100°C (212°F). But this only applies at standard atmospheric pressure, which is the average air pressure at sea level (101,325 Pascals or about 14.70 psi). To explain this, let's look at why a liquid boils at all.

Liquid water is made up of a bunch of water molecules all jiggling about. The molecules are not completely free to move anywhere they want as they are all holding onto each other. They are not holding on to each other as tightly as molecules in ice, which is why liquid water can flow and slosh about. I said that the water molecules are all jiggling and this is due to the heat energy they have. More heat equals more jiggling, but not all molecules have exactly the same amount of heat energy. Most will have an average amount but some will have less and some will have more. When one of those juiced up, high-energy molecules is at the surface of the water it may be able to break free from its neighbours. It needs to overcome not only the clutches of the other water molecules but the molecules of gas above the liquid trying to push it back in. So far, so good, as this explains why a puddle of water will eventually dry up without boiling. It slowly evaporates as the high-energy molecules break free. Now, imagine what happens when you heat the water up. You put more and more energy into the molecules and they jiggle faster and faster. When you hit 100 °C (212 °F) at standard atmospheric pressure, most of the molecules now have enough energy to not only break free from their neighbours, but push past all of the molecules in the gas above the water. In fact, at this point you start to see bubbles spontaneously forming inside the liquid, which then grow as more liquid rushes into the gaseous form.

But now imagine that the pressure of the air above the water is higher. For a water molecule to escape the liquid it needs to push past more gas molecules, which makes it harder and requires more heat energy. So, the boiling point goes up. If you increase the pressure to twice standard atmospheric pressure (about 203,000 Pascals or 30 psi), the boiling point of water rises to 120 °C (248 °F). Which is precisely what happens inside a pressure cooker. As the water begins to boil at 100 °C (212 °F), the water vapour or gas produced has nowhere to escape to, so it pushes the pressure up. In turn, this increases the boiling point and you continue heating and the water begins to boil again, and the pressure goes up and so on. Eventually Papin's safety valve comes into play at about double atmospheric pressure and the pressure stabilizes.

So, the physics of pressure and boiling point give us an extra 20 °C (36 °F) for cooking with, which does not seem very impressive. Until, that is, we consider the Arrhenius equation from 1889 that says if you increase the temperature by 10 °C (18 °F) a chemical reaction will proceed at twice the rate. So, the temperature inside a pressure cooker will cook food about four times faster than plain old boiling. Which is why, with all that delicious science at play, the pressure cooker wins the geek award for fast cooking.

There are a few other benefits to pressure cookers: they use much less water and hence energy than an equivalent long, slow stewing time and they also have temperatures hot enough to produce delicious flavour molecules through an essential bit of kitchen chemistry called the Maillard reaction (*see* page 100).

Pressure cookers have, like Papin, been overlooked in the last few decades. They are undeniably quick and efficient, but suffer from being very heavy and awkward to store and perhaps ever so slightly terrifying when the steam pressure needs to be released. On top of that, the meteoric rise of the microwave means that, at the moment, the pressure cooker remains a fringe interest at best.

Adding air

There is one essential bit of kitchen gadgetry that has not only a long and noble culinary history but, unlike the pressure cooker, is also found in every kitchen. The utensil I am talking of is the humble egg whisk. It's one of those items that seems

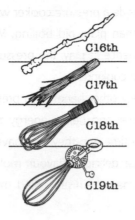

so fundamental you can't imagine a period when a cook would not have one to hand.

Such a time did exist and we can deduce its existence from documentary sources and the rise of the use of whisked egg in culinary instructions. The earliest recipe-book reference to whisked egg comes from the 1602 tome by Sir Hugh Platt called *Delightes for Ladies*, or to give it its full title, *Delightes for Ladies: to adorn their persons, tables, closets, and distillatories with beauties, banquets, perfumes and waters*. The book is full of hints and tips for use within the house and includes instructions on how to 'Break Whites of Eggs Speedily', which is Elizabethan code for beating eggs. However, the proposed method is to use either a knobbly stick or to repeatedly wring the eggs from a sponge. Neither of which sound very efficient. You may be thinking that a fork would be a better way to do this, but at the time this was an

THE SCIENCE OF FOOD

unknown item in northern European households. The fork first debuted in Great Britain in 1611 in an account of travels to Italy and the implement was still seen as an effete southern European affectation well into the late 1700s.

The results of squeezing egg from a sponge or whacking it with a stick barely constitutes whisked eggs. At best, you get a light froth. Then in 1651 we start to see recipes that are made with what must be whisked eggs. Sweetened, whisked-up egg was referred to as snow and plopped on top of a variety of different desserts. The assumption by culinary historians is that people must have moved on from knobbly sticks and sponges to whisks of some sort, probably made from twigs. There is even a recipe that suggests using cut and bruised apple twigs to whisk egg whites to impart an apple flavour. I'm not sure if the flavour of apple would come across but this is definitely the first use of a whisk. The thing is, you can't effectively whisk up egg whites using a single stick: you need a whole bunch made into a whisk, and this is where the science comes in.

The reason you can whisk egg whites is because the proteins in the albumin are denatured by the whisking and can more effectively hang on to bubbles of air. Whisks are very good at creating air bubbles. As each strand of the whisk, or twig if you are old school, moves through the liquid, it pulls air down and creates bubbles. If you do this in water the bubbles quickly rise and burst, but use a more viscous liquid and the bubbles hang around for a while. Something else is going on, though, as the bubbles in whisked egg can last for hours. As the whisk whips through the egg white, the physical action of the individual, bludgeoning whisk-stands breaks open the coiled-up egg proteins (see page 37 for how heat can do the same). This exposes the delicate inner workings of the protein strands – inner workings that are water-hating or hydrophobic. The now exposed, water-hating bits rush to the first place that contains no water, notably the air bubbles in the mixture. Every tiny bubble becomes surrounded with a sheath of broken down or denatured egg protein. These proteins quickly begin to bond to each other, forming a stable web of protein around every bubble.

At this point, a chef would describe your eggy mixture as soft peak. If you lift the whisk out of your mixing bowl any strands of foam will flop over, incapable of holding their own weight. But if you continue to whisk, the egg foam stiffens further to the stiff-peak stage. This is the point where you can infamously hold the bowl of egg over your head and it should stay there. As you continue to beat the egg and put more and more energy into the proteins, the bubbles get smaller and more numerous. This has a couple of effects: it makes the whisked egg look whiter and it reduces the amount of liquid between each bubble. The mess of denatured proteins around each bubble begin to tangle as they bump up against each other. As they do, they start to stick together and it is this that makes the whisked egg stiffen up. Since the proteins also start sticking to your bowl, it's also why you can do the bowl-over-your-head trick. The stiff-peak stage is the optimum for things like meringue that need to hold their shape in the oven.

However, continue to beat the egg and it all goes pear-shaped. You get what is known in the trade as dry peak. The foamy mass becomes almost crumbly and liquid begins to form at the bottom of the bowl. The problem you have here is that all those protein molecules are now pulling so tightly on each other that they start to squeeze the water in the mixture out from between the bubbles. The air bubbles in the foam can't now move around and the foam becomes partially set.

There are a few methods that have been developed over time to make the whisking process easier, with some proving more successful than others. While the rotary mechanical hand whisk will make the job of whisking less tiring, it never seems to deliver when it comes to beating eggs. The physical effort needed compared to the volume of foam generated does not seem to be worthwhile. You invariably end up resorting to a traditional balloon whisk and wondering why you bother to keep the mechanical version. The issue with the hand-cranked whisk is really that the beating action is all wrong. The recommended chef technique to beat egg whites is not to use a rotating action, but a vertical oval movement that lifts the egg and introduces lots of air. If you have an electric whisk, that will do a great job by virtue of the sheer speed of rotation, far faster than anything you can do by hand. But most chefs reject this as well as it's too easy to overbeat your eggs. The beaters are going so fast that it can be a matter of just moments between stiff-peak perfection and dry-peak disaster.

Your choice of bowl is also crucial, not least because the egg will increase in volume by eight times and you need big, extravagant motions of your whisk. But size is not the only thing that matter in your choice of bowl. If you can afford it, a copper bowl is universally regarded in the chef profession as best to use when beating your eggs. Tradition had it that you could not overbeat egg whites in a copper bowl. My immediate response when I saw this was that it sounded like a case of unverified tradition, but it turns out to have a significant effect that cooks had noticed way back in the eighteenth century. It wasn't until 1994 that we worked out why. As you beat your eggs in a copper bowl, a very tiny amount of copper dissolves into the mixture. Before you worry, it's well below the daily recommended dose for the metal. This copper then binds to chemically reactive sulphur groups on the denatured proteins. This prevents the formation of a particularly strong type of cross-link between proteins called sulphur bonds. The addition of copper effectively reduces the stickiness of the denatured proteins enough that they won't get too clingy and develop a dry-peak foam, but they are sticky enough to whip up to stiff peaks. It may slow down the whipping process but it definitely makes it easier. You get the same effect with bowls made of silver or gold, although I'll admit, those are harder to find.

Lacking the resources for a copper bowl, science can come to the rescue. Rather than adding copper to your egg whites, all you need do is add a little acid. A squeeze of lemon juice will do the job or, if you don't want to change the flavour, add a pinch of a dry powdered acid like cream of tartar (a common baking ingredient with a horrible proper name: potassium 2,3,4-trihydroxy-4-oxobutanoate). This acid will have the same effect as the copper by blocking the formation of sulphur bonds and it will make your egg whisking easier.

The addition of sugar is great for stabilizing your foam although it clearly has a major impact on the taste. The reason it helps is down to the increase in gloopiness or viscosity of the egg mixture. A more viscous liquid will hold on to bubbles more easily and allow the protein networks more time to form, so you need less elbow grease with the whisking. On the other hand, fat is the enemy of the egg whisk. If you have any fat in your egg-white mixture, it will massively reduce and even prevent the egg from being whipped up. The fat molecules effectively do the same thing as the protein molecules because they too are water-hating or hydrophobic. When a bubble forms, the fat will accumulate on the bubble's surface, competing with the protein for space. Consequently, the network of proteins never forms and your foam flops. So, how much fat will ruin your eggs? Folk wisdom proclaims that the slightest drop of egg yolk, which is packed with fat, will be your downfall. This misconception is easily tested and egg whites will whisk up fine even with a few drops of yolk in them. The other taboo you often see mentioned in cookbooks is that plastic bowls should be avoided. While it is true that oils and fats will cling to the plastic of the bowl and thus could pose a problem for the egg whisking, this is really just a sign that the bowl has not been cleaned properly with hot soapy water, which will strip the fat away from the plastic.

The egg whisk is a remarkable utensil that brings about a complicated change in an everyday food, turning it into something quite out of the ordinary. Armed with the science behind the transformation it creates, you should be better placed to whisk eggs like a pro.

Keeping cool

For all the wonders of the pressure cooker, the egg whisk and the sous vide gadgetry, none of these are as transformative as our ability to preserve food with cold and our dependence on electrically produced cooling. In 2015, the combined value of the chilled and frozen food sold in UK supermarkets was £18 billion, which is such a big number it is hard to get a grip on what that means. Put it this way: you could say that in 2015 every woman, man and child in the UK bought and presumably consumed £275 of frozen or chilled food. Of course, we throw away huge amounts of food and it is unlikely that a newborn infant would get through their allotted £275 worth, but you get the idea. It's a huge chunk of the food that is sold in the UK and it all relies on refrigeration.

There is one reason we refrigerate so much of our food chain and it all comes down to the Arrhenius equation, which you will remember from the pressure cooker explanation (*see* page 39). The simplest version of this is that the rate of a chemical reaction doubles for every increase of 10 °C (18 °F). By the same token, but in reverse, a drop of 10 °C halves the reaction rate. All of the things that can spoil your food rely on chemical reactions to change the molecules within the food. Reduce the temperature enough, the reactions slow down and the food stays in top condition for longer. Bacteria are the most common reason food spoils and these, like all living things, are essentially bags of chemical reactions. These conform to the Arrhenius equation just like all chemistry must, although due to the complex interactions within a living organism, the chemistry slows down even more quickly as the temperature drops. Which is why, if you freeze a lump of meat, it can last for much longer than the equation predicts. The safe storage time for meat that has a temperature of 21 °C (70 °F), or room temperature, is just two hours. After that the bacteria levels could be too high for safe consumption. Freeze the meat to -18 °C (0 °F), a drop of nearly 40 °C (70 °F) and the chemical reactions should slow to one-sixteenth of the rate (½ × ½ × ½ × ½). According to the maths, it should be safe to store the item for thirty-two hours at this temperature (2 hours × 16). Clearly biology has little respect for maths and you can store meat for much, much longer in a freezer since bacterial growth comes to a complete standstill at -18 °C (0 °F). Although, it's important to note it won't kill the bacteria.

People have been using artificial cooling for well over 3,000 years. Ice and snow would be harvested and stored underground or in special, thickly insulated ice houses and then used once the weather began to warm. However, most antique cultures just used their icy refrigeration to cool drinks. It is harder to find evidence for using the cold to preserve food. The closest we get in antiquity is the Persian dome-shaped *yakhchals* from about 400 BC. These were huge, 10-metre-tall (33 feet) cones made from a special mortar that used evaporative cooling to create ice in the winter. The ice was then stored through to the summer.

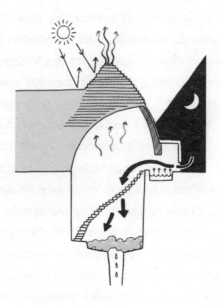

A Persian yakhchal.

We know the Persians used the ice to cool drinks and also to make *faloodeh*, a frozen dessert, but if they used the *yakhchal* to preserve food with cold, there doesn't appear to be any evidence. Many authors will claim they did, but this appears to be purely speculative. We take it for granted that if you have a chilled place, you are going to keep perishable food there so it keeps longer, but that is because we also take it for granted that we understand the science of food preservation and what causes food to perish (*see* page 150).

A more likely example of the first use of cold for food preservation is the native Inuit of what is now northern Canada. When Clarence Birdseye visited Newfoundland in around 1912, he observed Inuit flash-freezing the fish they had caught and then thawing it out later to eat. Mr Birdseye is often credited as the inventor of frozen food, but it's clear that the Inuit were using the frozen environment as a walk-in freezer long before.

For the rest of us, the chilled and frozen-food revolution began with the discovery of the vapour-compression cycle. At the heart of this is an observation made in 1755 by a Scottish chap named William Cullen. If you take a low-boiling-point liquid, such as ether, and put it at low pressure it evaporates and cools down, drawing in heat from its surroundings. It was something a lot of eighteenth-century scientists went on to investigate. When they soaked something in ether, the evaporation would cause significant cooling. Benjamin Franklin even had a go and commented in a letter to a friend that 'from this experiment, one may see the possibility of freezing a man to death on a warm summer's day'. Finally, in 1805, courtesy of American inventor Oliver Evans, we had a full description of how to create a cycle that pulls in heat during the evaporation step and gives it up with condensation. All that was needed now was to hitch the vapour-compression cycle to a machine that would pump the heat from inside a box, which became cold, to outside the box.

The first demonstrably workable and practical such device was the invention of Scotsman James Harrison, who had emigrated to Australia to work as a journalist. In 1856, he patented a machine that was used initially to make ice for the good burghers of Geelong, a city just 75 km (46 miles) to the west of Melbourne. Those canny Aussies were not slow to make further use of Harrison's cooler and soon had it installed in breweries and meat-packing companies. It was this second use that was to become Harrison's undoing. At the time, there was a major trade in beef from the US to the UK. The journey time was less than two weeks, the weather on the way predictably cold and, with a bit of ice on board, the carcasses would easily survive the voyage unspoiled. The boat journey to Australia was considerably longer and, in an effort to open up a competing trade, Harrison decided to send frozen beef to the UK. However, he was persuaded that it was too risky to install a refrigeration unit on a ship and instead, in 1873, built an insulated ice room on board the sailing ship *Norfolk*. Hundreds of cattle carcasses were duly frozen solid and packed in ice made using his refrigeration system. Sadly, his calculations were wrong, or maybe it was hotter than expected on the voyage, and the ice thawed. History does not record what became of his cargo, but the first venture in frozen-food delivery was not a success.

The birth of the frozen-food industry and the whole attendant revolution in our consumption habits came just a few years later with an even more epic journey. On 15 February 1882, the good ship *Dunedin* set sail from New Zealand to England fitted with a steam-powered refrigeration unit. The system burnt two tonnes of coal a day but kept the entire hold below freezing throughout the long passage through the tropics. The journey was not without excitement: there were fires, broken crankshafts and Captain Whitson even developed hypothermia while working in the hold. However, on 24 May the ship arrived in London with its perfectly frozen cargo of 4,331 mutton carcasses, 598 lamb carcasses, 22 pig carcasses, 250 kegs of butter, an unrecorded number of hare, pheasant, turkey and chicken, and 2,226 sheep tongues. Quite why you need that many sheep's tongues I don't know, but the age of refrigeration had well and truly begun.

Today the chain of logistics that runs from food producers through to our tables is rarely without a refrigerated step at some point. All your fresh vegetables, fruit and bags of salad need refrigeration to prevent degradation and retard ripening too soon. Then you have all of the frozen and chilled ready meals, dairy products, sliced meats, fish, juices and, harking right back to the beginning of refrigeration, cooled drinks. Most of the process that brings all this food to us is hidden; we just see the end result. However, a little while ago, I was fortunate enough to see just one section of how it all functions when I was involved in making a film. England has just three central hubs for frozen-food distribution. These massive facilities, known as ice cubes, are strategically placed around the country and take in frozen produce grown locally. This produce is stored and then sorted back onto refrigerated lorries destined for shops, not just regionally but all over the world. The scale of the operation is quite breathtaking. To sum up, within science, the results of any experiment are ultimately the end goal. However, the technical know-how and expertise to perform scientific experiments is clearly a vital component of understanding science as a whole. Sometimes how a result was achieved is as important as the result itself. In the same way, the science of food is as much about how we prepare the food as it is about the chemistry, biology and physics of the food itself. So, whether it is the blade of a knife, a fancy high-pressure or low-temperature cooking machine, or the humble egg whisk, the technology of our kitchens deserves a second look and there's more on the future of this technology later in the book (*see* page 163).

The Magic of Processed Food

Breakfast straight from a cannon

In the same way that there is more to kitchen gadgets than would appear, so many processed food products found on your supermarket shelves have some quite extraordinary and bizarre science buried inside them. Take, for example, the breakfast cereal known (depending where you are on the globe) as Honey Monster Puffs (formerly Sugar Puffs) or Honey Smacks. These are airy little morsels of puffed wholegrain wheat, glazed with sugar and fortified with vitamins. While I'm not a big fan personally, I do appreciate the effort that goes into making these.

First, you take wheat grain that has a moisture content of between 13 and 14 per cent. Then insert the grain into what is essentially a cannon with an airtight lid clamped on the end. You then place the cannon over a heat source and rotate the cannon along its length, tumbling and mixing the grains inside. As the grains are slowly heated, the pressure inside the cannon begins to climb; not only does the air inside expand, pushing the pressure up, but some of that moisture content turns to steam, which also increases the pressure. As the pressure is rising, something peculiar happens to the starch in the grains. In a process called gelatinization, it goes from being a hard, dry lump to a softer almost molten plastic-like substance. At a temperature of about 55°C (131°F), some of the water in the grain is absorbed into the microscopic granules of starch that make up the body of the wheat grain. The heat and the water break down the regular, semi-crystalline array of starch molecules, releasing them as long, squiggly spaghetti strands. Since these are free to slide over one another and are not locked in a crystal structure, the consistency of the starch changes to a jelly-like one. When the pressure inside the cannon reaches fourteen times atmospheric pressure (1.4 million Pascals or 205 psi), the clamp holding the lid on is whacked with a hammer, releasing the lid and all the pressure

with a bang. All the grains are flung violently out of the cannon and the sudden drop of pressure causes the hot water trapped in each grain to suddenly turn into steam, expand and puff up the gelatinized starch. As each grain cools, the starch sets and you end up with puffed wheat ready to be enrobed in sugar and vitamins for your breakfast-time delectation.

The process is known as gun puffing, since the puffed grain is literally shot out of the end of the pressure vessel. It's quite a spectacular process to watch. The noise alone is shocking enough: an almighty bang as the pressure vessel blows open followed by a deluge of puffed wheat. The same science underpins a variety of other puffed products, although many are not made with the gun. Puffed rice, another breakfast favourite, is made by putting partially cooked grains into a very, very hot oven at temperatures in excess of 250°C (482°F) and often up to 300°C (572°F). The rapid change of temperature boils water in the rice and puffs the gelatinized starch. It is also the same process that pops corn, although maize kernels have the advantage of each being surrounded by a tough seed coat that works just like a self-regulating pressure vessel. As the kernel is heated, the pressure inside builds until the seed coat splits and each kernel undergoes its own private decompression crisis, the expanding starch forming the popped corn so beloved of moviegoers around the world. Using these three methods, a whole host of grains can be puffed, not just wheat, rice and maize, but barley, oats, millet, sorghum and even quinoa, which isn't even a grain.

The science of steam-puffed gelatinized starch doesn't stop there. You don't even need to puff whole grains; a mixture of slightly dampened ground maize starch can be heated until it gelatinizes, then pressurized and squirted through a nozzle. As the hot starch exits the nozzle it depressurizes and expands to make a corn puff, which when liberally coated in powdered cheese becomes a cheesy Wotsit (in the UK), a Cheeto (US), a Kukure (India), a Nik Nak (originally South African) or a Twisty (Australia).

Making things thick

If puffed cereal is not your thing, the science of starch reaches further into so many of our processed foods. Once gelatinization has taken place (*see* page 59), the starch swells and becomes gooey as water is absorbed. If you keep raising the temperature a second process starts, confusingly called gelation. For something such as cornflour (maize starch), once the temperature hits 90 °C (194 °F) the starch begins gelation and it leaks some of the molecules from within the grains out into the surrounding water. Now, starch is superficially a fairly straightforward chemical made up of long chains of a sugar called glucose. Each glucose sugar molecule is itself made of a hexagonal ring containing five carbons and one oxygen. If you hook together a few hundred to a few thousand of these rings you end up with a very long, naturally curly, chain molecule known as amylose. Within the starch grain the amylose is packed in ordered arrays, each amylose molecule stuck to neighbouring ones in neat rows. However, once gelation starts and the amylose is out on its own in water, it's free to wiggle about and stick to anything it pleases in whatever disordered way it wants. All the amylose in the water begins to stick to other amylose molecules and knits itself into a complicated three-dimensional mesh. The water molecules now can't move as freely as before since this mesh gets in the way. If the water can't move so freely the mixture stops being so fluid and begins to thicken up. If the starch you added was cornflour mixed into meat juices from the Sunday roast, possibly with a drop of wine and a teaspoon of redcurrant jelly, you just thickened your delicious gravy.

Thickening a sauce with starch is not a particularly surprising thing as most of us have first-hand experience of this. However, the process outlined above is only scratching the surface of what can be done with long chains of sugars. It is of course more complicated than I have described. All starch is not the same and is never composed of just amylose. In fact, it's not even the major component of naturally occurring starch and only makes up 20 to 30 per cent of the weight. The most common molecule in a starch grain is called amylopectin, which is closely related to amylose, and also thickens sauces. But while amylose is a straightforward molecule that does a great job of creating a tangled mesh and thickening gravy, amylopectin is trickier. It has the same chemical structure, long chains of glucose, but rather than being one long straight line it has a fiendishly complicated branched and bushy structure. Even though it may have the same number of glucose subunits in a molecule, it's much smaller in comparison to amylose and consequently not so good at thickening sauces. If you imagine a load of long amylose molecules in water, they wave about, bump into each other, stick together and, hey presto, you are on your way to perfect gravy. Amylopectin, on the other hand, is much smaller, so doesn't bump into other molecules so often. To thicken with amylopectin you need much more of it and the result is less stable.

There is a whole field of food science called food rheology: the study of changes in consistency of liquids and gels. Food rheologists are dedicated to changing the consistency of processed foods mostly using chains of sugar molecules. You will have seen the products of food rheologists on the side of virtually all packaged foods. Most commonly it will be modified starch listed as an ingredient. Why, you may ask, do we need to modify it? Surely it does a perfectly good enough job as it is? Well, yes and no. If you are cooking at home or in a restaurant where the food is being consumed immediately, then plain old flour or maize starch does an admirable job when used as a thickener. If, however, you want to make a food product that can be packaged, transported, refrigerated, maybe frozen and stored on a shelf for possibly months then you have a much more challenging task. Good old reliable cornflour (maize starch) is rubbish if you are freezing your product. The thickened item will weep liquid from its surface as the chilled water is expelled from the amylose mesh, which for a commercial product will lead to it being rejected by the consumer. Also, if you are making a huge batch of a food that needs thickening, it is no good having an ingredient that does not behave the same way each time. The nature of starch is that the exact percentage of amylose to amylopectin can vary tremendously depending on the plant species used, the time of year harvested, the conditions the plants grew in and so on. So, the processed food industry needed a better solution.

One of the most commonly seen modified starches is called maltodextrin. It's made by taking amylose molecules – long unbranched chains of glucose – and snipping them up into shorter lengths of about ten glucose units. What you end up with is a very light and powdery substance that feels a little peculiar on the tongue. It dissolves incredibly easily, but aside from being a little bit sweet does not taste of anything. Which is true of most pure, modified starch products; they don't taste of much but they do display a range of interesting properties for the food rheologist to play with. Maltodextrin, for example, is very good at soaking up fats. You can use it to turn a liquid fat into a dry powder, without appreciably altering the flavour. Added to fried snack foods, it wicks up the residual oil so the product is not greasy to the touch or in the mouth. Since it is essentially flavour neutral and extremely soluble in water, it can be used to bulk-up liquid products, increasing viscosity without sweetness. It's for this reason that it is added to products like salad dressings. It's also a very stable product that won't change or spoil over time, making it an all-round champion for use in processed foods.

While modified starch is one of the main additives to processed foods, there are other long-chain molecules made from sugars that come from different sources. A particularly popular one is called carrageenan and you will see this on the ingredients list of many processed dairy products, but also on shampoo, non-dairy milks, processed sliced meats and even beer. The carrageenan molecule gets its name from carrageen moss, a type of red seaweed that is found growing on rocky

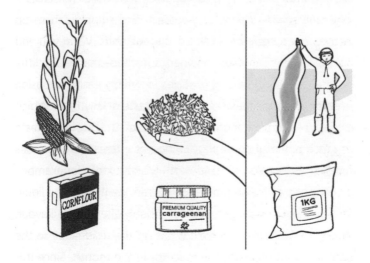

Long-chain molecules.

North Atlantic shores in both Europe and North America. It's a fairly common type of algae and has been used for centuries in traditional cooking from Trinidad to Ireland. If you boil up fronds of carrageen moss in milk, for example, when it cools it sets into a soft jelly a bit like a pannacotta or a blancmange. Although possibly with a slight residual taste of the sea.

The active ingredient from the seaweed, carrageenan is still produced by collecting seaweed and boiling it up in water. The resulting solution is then purified and has been extensively studied by food scientists. Its structure is superficially very similar to that of amylose, a long chain of hexagonal-ring sugar units. However, the devil is in the detail, and in this case the detail is all about a couple of bonds between carbon and oxygen atoms in each sugar subunit that stick up, instead of down, relative to the rest of the chain. If you compare amylose with carrageenan, the chemical formula is identical and the atoms are all connected together in the same order. The only difference is that in a couple of places the molecules are mirror images of each other. It's these differences that put the two molecules into different foods. One of the problems with amylose is that if you add it to dairy products, it reacts with some of the proteins and turns it all a bit slimy. Carrageenan, on the other hand, does not suffer this problem, which is why it's found in so many dairy products. It is also rather good at inhibiting ice-crystal formation, which is why you find carrageenan is used not only to thicken milkshakes and the like, but also in ice creams.

If these are your basic starch-like thickeners, there are also a few freakier ones. Take methylcellulose, for example. This is a synthetic additive made by taking cellulose, a halfway house between amylose and carrageenan, and adding a few extra carbon atoms along the chain. The result is a molecule that makes a gel, just like amylose and carrageenan, but this gel sets when hot and liquifies when cold, which is completely back to front. This stuff does not get much use in the food industry as its peculiar properties are not often needed. But just consider this for a moment: imagine you are making a deep-filled, mini apple pie. If you do this at home there is a risk that you will suffer from what is known as bake-out. The filling you load into your pastry case will be a satisfying thick consistency because the apples contain a lot of pectin, which is a very complicated mix of different chains of sugars that makes a gel just like amylose and carrageenan. As the pastry cooks, the filling gets hot, melts and turns runny. When it begins to bubble up, the runny filling leaks out of the pie, making a gummy mess on the pie tin and the floor of your oven. You have just suffered bake-out. If, however, you had thickened your pie filling with methylcellulose, as it heats up it would have become firmer and firmer. If the apple filling is not liquid, then it can't bubble out of the pie and bake-out is averted.

The last thickening ingredient I wanted to mention is possibly the most bizarre. Alginate is, as its name suggests, derived from algae or, to be precise, brown seaweeds such as kelp. Just as with carrageenan, the basic process is to boil up your seaweed in water, fish out the chunky bits and dry the result down to yield a white powder. The chemical structure of this elongated spaghetti molecule will sound remarkably familiar: a long chain of hexagonal sugar molecules. Once again, it is the way this molecule is organized and also the side groups of atoms attached to the hexagons that make all the difference. In this case, we have what are known as ester groups added that have the effect of making the molecule negatively charged when in water. While alginate does form a gel a bit like carrageenan and amylose, its secret superpower comes when you add calcium to a dilute mixture. If you take a mixture of calcium in water, it has two positive charges on each atom. Since the alginate has negative charges, the calcium grabs on to and sticks to these long-chain molecules. But, and this is the clever bit, since calcium has not one but two positive charges, it can grab on to two alginate molecules. When this happens, the molecules are cross-linked and set into a gel. So, to turn a dilute, cold solution of alginate into a gel, all you have to do is add some calcium.

Molecular gastronomists, or chefs with a penchant for playing with science, use this property of alginate to do some odd things. For example, make up a delicious, tart and fruity thin purée of plums and mix into this a little bit of powdered alginate (about one part in 200). Now get a bowl of dilute calcium chloride ready. Using a syringe, drip tiny blobs of your plum sauce into the calcium mixture. As the plum with alginate droplet lands in the calcium solution and begins to sink, the alginate instantly starts to turn from a liquid to a gel. It forms a thin, jelly-like layer around your tiny droplet of plummy goodness. You then fish the peppercorn-sized droplet out of the calcium water bath and rinse it in clean water, and you have made what is basically plum-flavoured caviar. As you bite into the fake caviar, each tiny blob yields a plum explosion in your mouth. Experimental chefs have used this to create all sorts of crazy food items. How does orange liqueur caviar dropped into a glass of champagne sound? It does not have to be tiny droplets. You can use alginate to make long noodles that liquify when you bite them or big blobs that look solid but gush out a sauce when cut into.

While all of this sounds exciting and different, you are probably thinking it has little to do with ordinary supermarket processed food, but think again. You will find alginate used in many places. As with carrageenan, it's added to a lot of dairy products like ice cream because, even without all the calcium excitement, it's a good, all-round thickener. And have you ever eaten a stuffed olive? If you look at the ingredients list for the pimento stuffing, it often includes alginate. It turns out that there are two ways to stuff an olive. One way involves laboriously poking tiny bits of pimento into an olive. The other way uses alginate. For this method, you first remove the olive stone, then inject a thick pimento and alginate purée into the resulting hole and drop the olive into a calcium-containing water bath. The calcium sets the pimento gel to the same consistency as whole pimento and there you have it, stuffed olives without the fiddle. They use the same method for garlic-

and anchovy-stuffed olives and, if you look closely on the labels, it usually says the olives are stuffed with a paste or purée. Does it make a difference to the taste or the mouth feel? It's hard to tell the difference unless you go to the effort of dissecting the stuffing and checking its internal structure.

Much of the science of processed food comes back, time and again, to the texture of the product and how to optimize this for the sometimes conflicting desires of not only the consumer, but the producers and the people tasked with transporting the product. Getting this balancing act right requires an armoury of clever chemicals, such as amylose, methylcellulose and carrageenan that can thicken or thin the consistency of the product at just the right time or temperature. With very few exceptions, all of these products are derived from the same basic chemistry: long chains of hexagonal sugars tangling together to make a gel.

When bread meets science

Starch and its many derivatives are not the only remarkable long-chain spaghetti molecules that play a vital role in our food production. If you are going to make bread, using the fermentation of sugars by yeast to create a light and fluffy loaf, you will also need some gluten in the mixture.

When you mix wheat flour with water to make dough, a number of things happen. Broken open starch granules absorb water, which releases enzymes that begin to digest the exposed starch. This in turn creates sugars that the yeast in your dough feeds off to make carbon dioxide gas, which makes the dough begin to rise. This is a crucial part of baking a loaf, but equally important is that two types of protein in the flour, called glutenin and gliadin, also absorb water. When this happens, the glutenin uncoils itself into long wiggly molecules, on to which the gliadin then binds. The result is a protein complex called gluten. Once the gluten has formed it quickly turns into a network of tangled strands as the gluten complex molecules stick to each other. It is this that gives bread dough its elasticity and spring.

Traditionally, at this point the sticky mass of wet starch and gluten network is kneaded for a good ten minutes or so. The whole purpose of kneading is to untangle the gluten network and help it become stronger. As the dough is stretched, the strands of gluten proteins are gradually lined up with each other – a bit like a tangled mass of hair being combed out. With the gluten proteins lined up, they are brought closer to each other and more chemical bonds form between them, making the gluten network stronger and stronger. The result is a silky smooth and elastic dough. You can see this happening if you make your own bread. Initially, if you pinch and then pull on the dough a little morsel breaks off and comes free. After sufficient kneading, the gluten network is formed and if you pinch and pull, the dough stretches and will even spring back if you don't pull too far. Now, when you leave the dough to rise, the elasticity of the gluten traps all the bubbles of carbon dioxide produced by the yeast, which when baked gives you light and fluffy bread. This process is one of the oldest known food-preparation techniques. It seems that humans have been making risen bread products for at least 10,000 years and probably much longer. There were no significant developments in the process for millennia. Until, that is, in 1961 when the Chorleywood bread process was invented.

Here in the UK we have a problem when it comes to growing wheat, the raw material needed to make bread flour. It's just not cold enough in our winters. Pretty much the entire of the UK is bathed in the warm Gulf Stream current that flows from the equator in the Atlantic Ocean. Our winters are not particularly harsh, especially not in the areas most suitable for growing wheat. Because of this, almost all of the wheat grown in the UK has a low protein content, generally only about 10 per cent of its weight. Unfortunately, that is not enough gluten to make good bread. For that you need at least 13 per cent protein, and ideally a bit more, in your flour.

Different varieties of wheat produce different quantities of gluten in their seeds. But all of the high-gluten varieties need a period of deep cold to get the best protein content – a period of deep cold that the UK just doesn't get. This process is called vernalization, which was first demonstrated and coined by the extremely controversial Russian scientist Trofim Lysenko. While his later work on eugenics has been thoroughly discredited, his earlier ideas about plants remain very much in the textbooks. Many different plants need vernalization, and the wheat varieties that do are known as winter wheat. In the northern hemisphere, the seed is usually sown between September and November. The seed germinates and grows into small wheat plants before the winter descends in earnest. Winter wheat needs a period of at least thirty days at between 0 and 5°C (32 to 41°F). The wheat will happily survive colder temperatures and lives on under a blanket of snow until the spring when it once again starts growing. Without vernalization, the wheat crop is smaller and, significantly for bread manufacture, does not contain enough gluten to make bread.

Winter wheat that has been successfully vernalized produces hard wheat, which in turn makes for strong flour. Conversely, almost all wheat grown in the UK is soft and only useful for pastry flour. To make bread in the UK, it was commonplace to import wheat, often from Canada where the winters are cold and wheat is hard.

This is where the Chorleywood bread process enters our story. Chorleywood is an unprepossessing village to the north-west of London, just outside the orbital M25 motorway. Since the Second World War it was also the home to the British Baking Industries Research Association laboratories. In 1961, scientists in these laboratories worked out how to make bread that didn't need strong flour and could be made perfectly well with soft British flour. Since then the use of the Chorleywood bread process has spread around the globe, and now around 80 per cent of bread in the UK is made this way. The key to the process is that you can turn the limited amount of gluten in soft flour into an effective gluten network, so long as you knead the bread incredibly violently for a short period of time. This is not a process that can be achieved at home as the machines required for the kneading are huge. The Chorleywood process is only suitable for very large batch baking. The intense kneading puts so much energy into the bread that it helps overcome any reluctance on the part of the gluten molecules to stick together to make an elastic network.

There are a couple of downsides to all this frantic kneading. Firstly, the dough gets really hot, which can interfere with the action of the yeast. To prevent this, you need to keep the sealed kneading drum jacketed in iced water. The other issue is that all the extra folding of the dough adds too many air bubbles, which give you a final bread with great holes in it. To stop these bubbles hanging around, the whole kneading drum is also placed in a partial vacuum. When the kneading is done, and it only takes a couple of minutes, the air pressure is returned to its normal, higher level and any air bubbles that formed are squashed flat.

So, the Chorleywood bread process allows the UK to make its bread with less imported flour. Clearly this has a consequence for the environmental impact of the manufacture as the food miles are significantly lower. An unintentional benefit of the process is that it takes much less time to make a loaf of bread this way. A traditionally made loaf may take up to twelve hours to produce. Using the Chorleywood bread process this can be cut to three and a half hours. This may not seem that significant, but in the UK alone we consume over 5 million loaves of bread a day.

There is one other process that has changed bread manufacture and is worth mentioning at this point. Like the Chorleywood bread process, it not only produces a loaf of uniform and fine-crumbed bread, but it departs radically from the traditional method of making bread. In Japan, bread products were introduced to the country by Portuguese traders in the seventeenth century. One of the first products brought over was traditional Portuguese sweet bread, or *pao duce*, which has a very fine and soft texture. Gradually bread found its way into more Japanese cooking, including taking the *pao duce*, crumbling it into fine crumbs and using it to enrobe other foods that were then deep fried to give a crispy coating. Legend has it that during the Second World War, Japanese soldiers in the field had no way to cook bread, until some bright spark decided to use a tank battery. The result was panko bread and this is, to my knowledge, the only food anywhere in the world cooked using direct application of electricity. These days huge banks of square steel tubs about 30 centimetres (12 inches) across and 10 centimetres (4 inches) deep have a blob of dough pushed into each of them. The tubs are arranged in large arrays that are then connected to a high voltage and high current flow of electricity. As the electricity flows directly through the dough, the movement of electrons creates heat through friction, which in turn cooks the bread. The result is a block of bread with no crust and a very fine, uniformly bubbly texture. Within Japan, some panko loaves are chopped up into sandwich-sized slices but the majority is air-dried for about eighteen hours before it is carefully shredded into sliver-shaped panko breadcrumbs. If you have ever eaten any Japanese food coated in crispy breadcrumbs they will have been made with this unique process.

People often worry that supermarket bread is somehow less good for you than an equivalent artisanal loaf, or that it is processed in a way that a homemade loaf is not. While it is true that a wholegrain loaf may contain more fibre, if you make bread from white flour it's going to have pretty much the same nutritional value no matter how it is made. What's more, in a recent study, scientists working at the Weizmann Institute in Israel showed that, aside from the fibre content, the type of bread you eat makes no difference. They had two groups of test subjects eat just four slices of bread for their breakfast over the course of a week. One group were given sliced white Chorleywood bread and the other a wholemeal sourdough bread. The results were interesting mostly because of what they didn't show. Samples from the test subjects were taken and analyzed to find out how quickly the bread was converted into glucose in the blood. They also looked at the gut microbiota (see page 137) to see if the food was having any long-term effect on the subjects' digestive system. In both cases, there was no significant difference between the wholemeal sourdough and the white Chorleywood bread. In fact, the greatest differences were between people. Some test subjects had a big spike in blood glucose levels after eating Chorleywood bread and some when eating the sourdough bread. The biggest factor that determines how your body responds to eating bread is not what kind of bread you eat, but your specific genetic make-up and the range of bacteria living in your gut. Perhaps this should not come as a big surprise as all bread is made predominantly from one ingredient, flour, and all flour is itself predominantly made of starch.

The Chorleywood bread process and, in Japan at least, the production of panko bread has revolutionized the baking industry and allowed us to make uniform loaves that minimize the food miles on our bread. So whether it is high-speed, high-energy kneading or the application of direct electrical current, it just goes to show that despite bread being the world's first processed food, the twentieth century turned up a few new surprises along the way.

Food in an instant

I first encountered instant mashed potatoes way back in about 1989 when I was on a walking trip in the Yorkshire Dales. While the cooking facilities where we were staying were ample, there were very little options to buy actual food. I resorted to purchasing a packet of instant mashed potato. I had never tried the stuff before and, I'll be honest, the experience was not great, although this may have been to do with what else I added to the mash to try to turn it into a meal. Apparently, canned corned beef and instant mash is not a good combo. However, what did astound me was how you could go from a small mound of dry, flaky stuff to a panful of mashed potato in mere moments. The transformation takes about ten or twenty seconds at the most. Instant mash, with its nigh on magical properties, is at its heart a matter of simple dehydration. Except, of course, that it's not that simple.

If you take some potatoes, mash them up and then try to dry them out as a big lump, you run into all manner of problems. You can't just put them in a pan on the stove and heat them up. To maintain the consistency of mashed potato you need to be careful not to overheat them. If you do, the starch grains begin to break down, change colour and give the potato a burnt caramel flavour. Equally, you must be careful not to beat them too vigorously as this will gelatinize the starch and you end up with a vile gloopy and glue-like mash. On top of this, the dried potato must be at least 95 per cent free of water. If it isn't, not only will it quickly spoil when stored, but it becomes more difficult to rehydrate without lumps. So, you can see that drying out mash is a more difficult process than it perhaps seems on the surface.

The industrial secret to drying out mash is the rotating drum dryer. To be honest, it's not much of a secret, it's just not very well advertised. You start with a big pile of perfectly mashed potato. Then, in the basic version, this is plopped on to a slowly rotating and heated drum. As the drum rotates, a half-millimetre smear of mashed potato is pasted on to it. The surfaces of rotating drum dryers are heated internally with pressurized steam, so they run at above 100°C (212°F) and up to 200°C (392°F). Water in the mash smeared on the surface quickly evaporates and by the time the drum has performed about a half rotation, in just a few seconds, the mash has dried. The resulting wafer-thin sheet of dry mash is scraped off with a long blade, broken into small pieces and you have instant mash. On bigger models the drum is over a metre in diameter (over 3 feet) and up to 5 metres long (16 feet). They are fitted with lots of smaller rollers that smooth and gradually thin the mash from a centimetre-thick layer (about half an inch) to the dried-out, wafer-thin sheet. Even a basic small model can produce a couple of kilograms (about 4½ lbs) of instant potato flakes in an hour.

There are two reasons that the food-processing industry uses drum dryers. Firstly, they are really good at drying out thick, gloopy mixtures like mash or fruit purées. The mixture

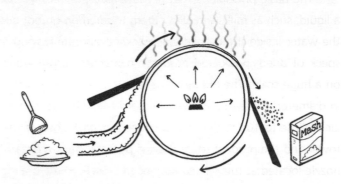

being dried will readily stick to the drum without falling off and comes away cleanly, allowing for a continuous process. The second advantage is that the result is flaked and not powdered. When it comes to rehydrating something as instant as instant mash, powders tend to result in lumps if you are not very careful with your stirring. The flakes don't suffer this as much since, by virtue of their shape, they don't pack down so neatly and water can flow into and around a pile of flakes easier than a pile of powder.

While the drum is ideal for mash, there are many instant foods where it doesn't work. Take powdered milk: not only is the starting material clearly too liquid to work on a drum, but the temperatures used would cook the milk and massively change its taste. For powdered milk, you need a spray dryer. At its heart, this is just a machine for removing water but it does it in such a way that the drying temperature is kept as low as possible, much lower than is used on a drum dryer.

The basic principle is that if you create tiny droplets of a liquid, such as milk, and allow them to fall through hot air, the water inside the droplet will quickly evaporate leaving a speck of dried, powdered product. In practice, this is done on a huge scale. The big industrial spray dryers are 6 metres in diameter (20 feet) and over 30 metres tall (100 feet). Hot air at close to 100°C (212°F) is blown in near the top of the tower and the product to be dried is pumped through a fine nozzle located at the very summit of the tower. There is a lot of careful science that goes into the nozzle on a spray dryer as it is this that controls the particle size. Too small and you get dust, too big and the droplet won't dry out before it falls to the bottom of the tower. At the bottom, the dry powder and all the hot air and steam are drawn into a whirling cyclone chamber, a bit like on those funky bagless vacuum cleaners. The powder drops to the bottom and the steam and air are whisked away.

For some spray-dried products, that is the end of the process. The powder is packaged up and sold to you and me. Things like powdered soup and powdered stock are left at this simple, fine-powder stage. The most common powdered foods though, like milk and coffee, go through a second step. A little water is sprayed back onto the powder, which is then jiggled about to make it stick together in little, loose clumps. This is mostly a cosmetic step as, apparently, we the consumers place a higher value on little crunchy granules rather than a uniformly fine powder.

However, neither the drum nor the spray dryer are the truly science-geek way to dry things out. The problem with both methods is this: to dry out food you need to add heat. You could apply a very low heat, say 50 °C (122 °F), for a long time, but you risk the food spoiling before it dries out properly. Alternatively, and this is what drum and spray dryers do, you apply much higher heat for a very short period of time. The problem is that heat changes chemicals and in particular flavour chemicals, even over short exposures to high temperatures. It's not such a problem with powdered soup and mashed potato, as these contain relatively few delicate flavour chemicals. Something like coffee, on the other hand, is packed with subtle aroma molecules. So, industrial food processing has come up with cunning ways to remove water without applying heat.

All chemical reactions go faster as temperatures go up; it's that pesky Arrhenius equation again (see page 43 and page 51). So, while you can drive off the water with high temperature you are always going to have chemical reactions happening that alter the flavour and texture. It's why getting the water to evaporate really quickly is so important. Drum and spray drying represent our best efforts to minimize chemical changes while maximizing water evaporation. Thankfully, with the appliance of science, you can boil water away at much lower temperatures.

The speed of the pressure cooker (see page 39) relies on an increase in pressure, making water boil at a higher temperature. By the same principle, if you decrease the pressure, water boils at a lower temperature. This is something that tea-loving mountain climbers suffer from. When you climb to the top of a mountain, the pressure drops and the temperature at which your water boils also drops and you can't make a decent cup of tea. As an Englishman, this is of utmost importance to me, and I can assure you that water used for brewing tea should be at, or very close to, 100 °C (212 °F). At the top of a mountain just 3,000 metres tall (10,000 feet), the boiling point of water is below 90 °C (194 °F) and a disaster for brewing tea.

The thing is, while the boiling point of water drops away as the pressure drops, the freezing point of water essentially does not change. When you get to a very low pressure (six thousandths of atmospheric pressure, 611 Pascals or 0.09 psi), the boiling point of water drops to 0 °C (32 °F), which is the same as the freezing point. It's called the triple point of water because water can exist at this low temperature and pressure as solid, liquid or gas. Below this triple-point pressure, you can't have liquid water. Since the boiling point and freezing point are now the same temperature, as you warm up a block of ice at, say, two-thousandths of atmospheric pressure (200 Pascals or 0.03 psi), it does not melt; instead, it goes straight to boiling and turns into water vapour. It's a process known as sublimation and, at this pressure, it happens at -20 °C (-4 °F).

It's at this point that we can start to make freeze-dried coffee. The flavours in coffee come from a complicated mix of very delicate molecules that don't handle heat at all well. This is why spray-dried instant coffee does not really taste much like freshly brewed coffee. That complicated suite of aromas is mangled by the heat. So rather than spray drying, consider this. Make yourself a very strong batch of coffee, pour it into a shallow tray and freeze it down in an industrial freezer at -25°C (-13°F). Now, without letting it warm up, break it into lots of tiny little pieces and put it into a vacuum chamber. Then pump the air out of the chamber until the pressure is only two-thousandths of atmospheric pressure (200 Pascals or 0.03 psi). Finally, allow the frozen coffee to warm up ever so slightly to -20°C (-4°F). At this temperature and at this low, low pressure, the water in the frozen coffee begins to sublime, turning into water vapour, and is pumped away. The frozen nuggets of coffee flavour dissolved in water turn straight into dried-out nuggets of coffee flavour, and the temperature never went above -20°C (-4°F). All those flavour molecules in the coffee are unaffected by heat and remain in the freeze-dried granule. Or at least they mostly remain in the granule. As the water is subliming from the coffee granule, so are some of the flavour molecules. There is nothing that can be done to stop this. So, what the producers do instead is trap the flavour molecules, turn them back into a liquid and spray them onto the freeze-dried granules of instant coffee, which goes at least part way to keeping the end result tasting a bit like real coffee.

Taking water out of food seems like such a trivial thing to do and yet it can create some of the most remarkable processed foods you are likely to find on your supermarket shelves. Not only that, but in our search for better and better ways to dehydrate food, we have invented some incredibly ingenious machines and processes that get used for all manner of things. Take the new rotating drum dryers that work inside a vacuum chamber and are being used within the pharmaceutical industry. Or there is the spray dryer that I recently came across that creates a superfine and uniform powder used to prepare soil samples for forensic analysis. The Scottish forensic soil lab who developed it can pinpoint where the mud on your boot came from down to around a hundred metres (about 300 feet). So long as it's in Scotland. As for freeze drying, well it's being used for more than just coffee and soft fruit such as strawberries and raspberries. It is possible to freeze dry vaccines, perishable drugs, blood plasma and enzymes, all of which would normally need to be kept refrigerated, but once freeze-dried can be stored at room temperature as a powder (although *see* page 195 for the latest on keeping liquid vaccines refrigerated without a power source). Freeze drying can even be used for recovering water-damaged old books and archaeological finds.

Making the impossible mixture

Oil and water do not mix. At least that is the well-worn aphorism. If you take a load of vegetable oil and dump it into

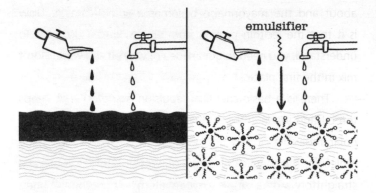

a glass full of watery vinegar, the oil bounces up to the surface and floats to the top without mixing. Try to mix it, maybe by giving it a whisk, and as soon as you stop, the oil droplets float to the top and you are back where you started.

And yet these are the main ingredients for mayonnaise. How then does the oil and vinegar in the mayonnaise manage to join together and stabilize? The secret to this impossible mixture is the addition of egg yolks that contain a very special substance called lecithin.

Traditionally you mix the egg yolk and the lecithin it contains with your vinegar and then whisk like crazy as you pour in a thin stream of oil. The oil is gradually incorporated into the liquid and forms tiny, stabilized droplets. The mixture thickens as you beat the oil into smaller and smaller droplets. In the end, you have about four times as much oil as you have of the watery part of the mixture. The water is spread thinly around each oil droplet, so thin that the droplets can't move about and the mayonnaise becomes a smooth paste. How is it that the lecithin protein from egg yolks can do this? To understand, you need to get an idea of why oil and water don't mix in the first place.

There is a basic chemical repulsion going on that keeps the oil and water from mixing. Water is what is known as a polar molecule: it has charged poles or regions within the molecule. The chemical structure of water seems on the surface straightforward: a single oxygen atom with two hydrogen atoms stuck to it. But there is more to it than such a simple description belies. The oxygen atom within a molecule of water has a slight electrical negative charge and the hydrogen atoms a correspondingly small electrical positive charge. Since opposite charges attract, all the water molecules in the liquid are busy jiggling around being slightly attracted to each other. If you add another polar molecule to the water, like alcohol for example (technically I'm talking about ethanol), it readily gets in among all the polar water molecules and the two substances mix together. The alcohol in this case is called a hydrophilic substance, which just means water (*hydro*) loving (*philus*).

Oil molecules, on the other hand, are non-polar. They do not have charged poles or any electrical charges anywhere in them. The structure of oil is much more complex than water and can vary depending on the source of the oil. However, all oils share a common basic structure, a series of three long carbon chains joined together at the top. The nature of the vegetable oil depends on the three long carbon chains – each of these can be different and every combination yields different properties – but all of them are non-polar and have no electrical charges lurking anywhere in the molecule. While this makes no difference to the oil molecules, when they meet up with a polar molecule, like water, they don't get on. The water molecules won't let the oil join their polar molecule party. There is nothing drawing the polar and non-polar molecules together, so they stay apart as a whole. Oil is a water-hating or hydrophobic substance.

Oil and water don't mix unless you introduce something called an amphiphile. The thing is, molecules can be really big and it is entirely possible to have sections of a molecule that are hydrophilic (water loving with polar electrical charges) and sections of the same molecule that are hydrophobic (water-hating with no electrical charges). These molecules are called amphiphiles (from the Greek *amphis*, meaning both) as they are happy mingling with both water and oil. But the clever bit happens when you mix oil with water in the presence of just a smidgeon of an amphiphile. When an amphiphile is in a mixture containing both water and oil, it will naturally make its way to the interface between the two. The hydrophobic parts of the amphiphile will sit inside the oil layer and the hydrophilic, polar portions will sit in the water layer. The amphiphile forms a layer, a single molecule thick, all along where the oil meets water.

So far so good, but now whisk up your oil and water non-mixture. You create a multitude of tiny blobs of oil floating in the water, each surrounded by a mass of water. But rather than the droplets sticking together, coalescing and turning quickly back into a layer of oil, the droplets persist. The amphiphile coating on each droplet hangs onto the water around it, preventing neighbouring droplets joining together. Keep beating the oil and water mixture and the droplets get smaller and smaller. The mixture will change colour to become a creamy white as all those tiny droplets start interfering with any light passing through the mixture. Eventually, there are so many droplets, each only a few thousandths of a millimetre across (a few tens of one-thousandths of one inch), and the water in your mix is spread so thin that the droplets can't now move around and past each other. At which point the mixture begins to thicken and ceases to act like a liquid, even though it is made of two liquids mixed together. And that is how you get the vinegar and the oil in mayonnaise to mix together. A smidgeon of amphiphile is all it takes to perform the impossible and mix oil with water. The resulting mixture is called an emulsion.

Once you start looking for emulsions in food, they turn up all over the place. Cream, for example, is a classic emulsion of fat suspended in liquid. In this case, the amphiphile emulsifier is the milk protein called casein. Like all proteins, casein is a long-chain molecule. Unlike many proteins, though, casein is not just made up of hydrophilic polar sections; it has hydrophobic, water-hating, sections too, which makes it a great emulsifier. Butter is also an emulsion, with casein as the emulsifier, but in this case it's a back-to-front or inverted emulsion. The droplets in butter contain the water and these are surrounded by the fat.

THE SCIENCE OF FOOD

The science of emulsions and emulsifiers is a crucial weapon in the arsenal of food manufacturers. You will find emulsifiers shown on the ingredients list of many of your favourite food products and not just mayonnaise. By creating an emulsion of fat and water in a food product, it allows you to do a few clever things. Firstly, you can change the texture of the product without adding a thickening agent (*see* page 61) and without adding more fat than is already there. You can change what is known as the mouth feel from something runny to a smooth and velvety texture. In fact, you can often make something taste more unctuous and creamy even with a reduced amount of oil, making it an ideal way to produce low-fat products. For foods that contain higher amounts of fat, like cakes and biscuits, adding emulsifiers helps prevent the oil from separating out of the product, especially when it's stored at room temperature on a shelf in a shop. Probably the best example of both these effects of an emulsifier at work is in processed cheese.

Now, while I realize that not everyone is a fan of processed cheese, it does solve a problem within the food industry. Imagine you are making a cheeseburger for yourself. You place a slice of hard cheese like cheddar on the burger, pop it in the bun, maybe add some relish, lettuce and a spot of mustard, and eat it. Delicious! But imagine you are making hundreds or thousands of burgers in a takeaway restaurant. Your burgers are not going to be eaten immediately. The cheese will sit on the hot burger for a while, maybe ten minutes, before consumption. If you take a piece of hard cheese and warm it for just a minute, the fat within the cheese begins to leak out and pool on the surface. If you use ordinary cheddar, your burgers will drip grease in a most unappetizing way that will not result in repeat customers. The issue is that cheese, while it is an emulsion, is not a very stable one. Just a small amount of heat and it begins to break down. The fat droplets fuse and the emulsion collapses, oozing grease, which is where processed cheese comes to the rescue.

A slice of processed cheese contains about 60 per cent or more ordinary cheese. The strength of the cheese you start with determines the final strength and flavour of your processed cheese. Want a full-bodied processed cheese? Then start with a grated mature cheddar and add to this some water and some whey powder. The water is there to allow us to create a better emulsion of the fats within our starting cheese, and the whey powder adds bulk and gives the final processed cheese a creamy texture. Whey powder is a mix of milk sugars and proteins (although not casein) which is made by drying out the leftover liquid when you make a traditional cheese in the first place. The only other important ingredient is what are collectively known as emulsifying salts. Somewhat confusingly, emulsifying salts are not emulsifiers themselves but instead are a mix of different types of phosphate. What is important about the phosphates is that they are very, very polar molecules, with lots of negative charges that are very keen on sticking to things with positive charges. And this is how they help emulsify cheese.

Inside any dairy product the emulsifier is the casein protein, but in cheese, especially mature ones, the casein gets broken into fragments. These fragments then stick to calcium that naturally occurs in milk, and when this happens they stop being very good emulsifiers. The calcium clogs up all the polar, water-loving, sections of the casein and it ceases to be a very good amphiphile. If you add emulsifying salts, chock full of phosphates, these pull all the calcium away from the casein and lock it away so it can no longer interfere with the formation of an emulsion. The casein is now free to do its work unimpeded.

The method for making processed cheese is fairly straightforward. You take all the ingredients, including the essential emulsifying salts, put them in a heated bowl and start vigorously mixing. As the cheese melts, the emulsifying salts do their work and the casein begins to allow the water and fat to properly emulsify. When I tried making my own processed cheese it was an almost magical bit of cooking. Nothing happened until I hit the right temperature of about 70 °C (158 °F). At which point, over the course of just a few seconds, the mixture went from a sloppy mess to smooth processed cheese. I then flattened my processed cheese between two baking trays to create a big, thin sheet perfect for slicing into squares. The same method is used but on an industrial scale to make the cheese squares used on burgers all over the world. By tweaking the exact amount of cheese, the maturity of the cheese used, the amount of water added and the mix of emulsifying salts, you can create a processed cheese of just the right melting point, which does not go too runny when molten and definitely has no oozing fat.

The processing of food by creating emulsions is an incredibly useful tool for food manufacturers. Whether it's diet foods, a specific melting point or just not leaking fat everywhere, emulsifiers let food technologists create amazing products tailor-made to our needs. Now, this may be controversial, but I suspect that things like processed cheese are given a bad press, not because of the way they taste or the ingredients used, but simply because they don't behave the way we expect them to. If you try cooking a traditional cheese-based recipe and substitute in processed cheese, it won't work. Designer processed cheese has its own behaviour when cooked that is usually tailored to one job alone: in the case of processed cheese squares, as the ideal thing to put on top of burgers.

Engineering for sweetness

The human being is conditioned by millennia of evolution to find certain foods appealing. It's the reason why we enjoy eating sweet, fatty or carbohydrate-laden foods. Our bodies and brains are hard-wired to find them enjoyable to eat. The reason for this is simple: these are all energy-rich foods and, from an evolutionary perspective, it is advantageous to seek out high-energy food. If an organism lives in an environment where food is scarce, or even just not very abundant, seeking out sweet or fatty foods is an advantage. They contain lots of energy and the organism will thrive. So, our liking of these foods comes from our deep evolutionary past. But there is a complication. We now find ourselves, at least in developed countries, with an abundance of cheap, energy-filled foods. If you allow yourself, you can eat as much sweet or fatty food as you want and satisfy our insatiable desire. Except, of course, you will put on weight. The maths is very simple, if the amount of energy you eat exceeds the amount of energy you expend in the day, the excess energy will be turned into fat. The problem is people are complicated and we don't just eat because our body needs energy and neither do we necessarily stop when we have had enough. The entire sugar-substitute industry is driven by this simple but far-reaching bit of science. We want a way to indulge our sweet tooth but without the energy that comes with regular sugar.

The first sugar substitute to be manufactured and sold as such was saccharin. It was an accidental discovery by Constantin Fahlberg and Ira Remsen, chemists working at the Johns Hopkins University in Baltimore in the United Sates. Fahlberg was an expert on sugar, the ordinary fattening kind, and was employed by the H. W. Perot Import Firm as an expert witness in a trial over some contaminated sugar imports. Fahlberg was tasked with testing the dodgy sugar for impurities. For this, he needed a workspace and it was arranged that he should use the facilities in Ira Remsen's laboratory, who worked in a similar organic chemistry field. Once Fahlberg had completed his analysis and was awaiting his opportunity to testify at the trial, he was allowed to carry on his own research in the laboratory. One fateful evening in 1878, Fahlberg returned home to join his wife for a meal that included some bread rolls. As Fahlberg ate the rolls he realized that they tasted remarkably sweet, a sweetness his wife could not taste. He quickly realized the sugary taste was coming from his own fingers and that he must have spilt something on his hands in the laboratory. The story goes that he raced back to the lab and tasted all the glassware on his bench until he found the culprit: an over-boiled beaker where three chemicals had accidentally reacted to create benzoic sulphimide or saccharin. Together with Remsen, he published papers on the new artificial sugar. As far as we can tell, Fahlberg took out patents for the manufacture of saccharin and set up a factory in Germany to make the stuff commercially, all without telling Remsen, and certainly without giving him any credit. The two were not on friendly terms after the discovery. What I find most perturbing about this story, though, is that not only did Fahlberg fail to wash his hands after working in a lab and before eating, but that he then went about licking all his glassware to discover the source of sweetness.

Following its discovery and commercial production, saccharin remained an obscure additive to food until the First World War, at which point sugar shortages saw a huge increase

THE SCIENCE OF FOOD

in production. We still weren't using it for its low energy, but just as an alternative to sugar. Until, that is, in 1958 when two products came onto the American market to cater for the growing trend in low-energy, or low-calorie, foods. The first was called Sweet'N Low and the little pink packets can still be found nestled in sugar bowls all over the world, although in some countries it no longer contains saccharin. The idea that you could sweeten your tea of coffee without adding energy or calories to the drink was in a way a tiny revolution in our thinking about food. The same year the Royal Crown Cola company introduced a new type of soft drink to its range. The Diet Rite cola drink was marketed, as its name implies, for the specific purpose of losing weight or maintaining a diet.

This was the start of the rise of the consumption of artificial sweeteners. Why, you may ask, do we need so many types of sweetener though? The truth is that most artificial sweeteners, while being sweet on the tongue, don't taste of sugar. Saccharine, for example, leaves an unpleasant bitter aftertaste as the molecule breaks in two after binding to the sugar receptors on your tongue. One of the now broken apart components is decidedly unpleasant to taste. To get around this it is usually mixed with another sweetener called cyclamate. This doesn't have a particularly convincing sugary taste either, but does mask the bitterness of saccharin. As an aside, cyclamate was discovered in 1937 by Michael Sveda when he picked up and smoked a cigarette he had accidentally dunked in spilt chemicals on his laboratory bench.

Once again, we have serendipity and lax health and safety to thank for his discovery.

Sweeteners are not without controversy and concerns have been raised over how safe they are to consume. Aspartame, which is used in a huge number of carbonated drinks, including Diet Coke and Diet Pepsi, was discovered by chemist James Schlatter in 1965 when he licked his finger (a pattern is emerging) and has been associated with numerous potential health issues including: seizures, migraines, panic attacks, weight gain, weight loss, attention deficit hyperactivity disorder, methanol poisoning, increase in appetite, changes in breast milk and causing cancer. However, the scientific literature and reviews of clinical studies do not show any significant causal links between consumption of aspartame and the potential health problems. There have been one or two studies that showed clinical effects, but there have been far, far more that showed none. This pattern has been repeated with many of the artificial sweeteners: oftentimes laboratory tests show toxic effects on rats, for example, but none in a human population. It should be mentioned that one historically famous sweetener did turn out to be very toxic; lead acetate or sugar of lead was used extensively as a sweetener by the Romans and throughout the Middle Ages. It's easy to make, was often used to sweeten wines and unsurprisingly gives rise to lead poisoning, which is a very nasty way to go indeed. Lead acetate aside, other sweeteners appear to be just that: sweet but without any energy when ingested.

Probably the biggest scientific argument right now in the field of artificial sweeteners is what impact eating them has on diet. Now, it seems like it should be simple. If you use aspartame or the newer sweeteners derived from the *Stevia rebaudiana* plant species, you eat less sugar and thus consume less energy. If over a prolonged period the amount of energy you eat is less than the amount you expend during the day you lose weight. Except there have been a number of studies that show the reverse. In fact, at the time of writing there have been ninety studies on this and twenty-eight (just under a third) show an increase in bodyweight. Subjects were fed either sugar-sweetened drinks or zero-energy sweeteners in their water and then allowed to eat as much or as little as they wanted. In some cases, the subjects with artificially sweetened drink ate more and got fatter. Which is strange and not at all what you might expect. The problem with these studies is that the subjects were all rats and not people.

Studies involving human beings are much, much harder to do. This is a basic problem with all nutritional studies. Not only are we all genetically very different but we all lead wildly different lives. Taken together this makes it hugely difficult to pick apart the effect eating something has on our overall health. The reason people do rat studies is that they can make all this variation disappear and do proper science with proper controls. Researchers can use genetically identical rats and dictate exactly what they all eat. There have been a few of what are known as cohort studies, where a bunch of people were fed sweetened drink, either containing sugar or an artificial sweetener, and monitored for changes in their eating habits. In some cases, people ate more, either long term or in the short term, when they were on the artificial sweetener. But just as many showed the reverse effect or no effect. The results were all over the place and when taken together were inconclusive.

What we can be sure of, however, is that if you eat too much sugar this is definitely linked to weight gain, and weight gain is one of the causes of cardio-vascular disease, diabetes and cancer. Reducing your sugar consumption is going to be good for you. Using artificial sweeteners may help people achieve this goal but, as always, moderation is probably wise.

The chemists who have pioneered the discovery of artificial sweeteners have provided us with a tool to reduce our intake of food energy should we wish. Admittedly it seems that mostly they did this by failing to wash their hands and smoking in the lab, but the results are there for us to see and taste in more hygienic ways should we choose. It turns out you can have your cake and eat it, at least as far as the sugar is concerned.

Critical Kitchen Chemistry

The king of kitchen chemistry

The aim of any chef working in a kitchen is to create a dish of food that tastes great. So far so obvious, but to do this, to create wonderful flavours, is to set out on a journey into chemical reactions that produce a dizzying array of molecules. There is one reaction, though, that stands head and shoulders above all others in this chemical extravaganza. The Maillard reaction is what puts the delicious into bread, cooked meat, coffee, soy sauce, beer, chocolate, popcorn, fried onions, cookies and so many wonderful-tasting foods. Chefs have been using it for thousands of years, but what was actually going on was only fully explained just over a century ago in 1912 by French physician and chemist Louis Maillard, working at the University of Paris.

However, before we delve into Maillard and his reactions, I should probably nail down what I mean by flavour. There are two ways in which we can taste a flavour. There are the usual sweet, sour, salt, bitter tastes we all learnt in school. On top of that we now have umami, the meaty or mouth-filling taste for which receptors were identified only relatively recently (*see also* oleogustus on page 107). All five of these flavours are detected on our tongues by specific receptors in our taste buds. I like to think of these as the big building blocks of how something tastes. These are the dominant notes that a chef assembles.

On top of this, though, are the subtleties that give the range of flavours we can enjoy in food. For these flavours, we don't rely on our tongues but on our nose and sense of smell. Such flavours are the icing on the cake and all arise from what are known as volatile compounds in our food. All this means is that there are molecules within the food that readily and quickly turn from liquids to gasses when they are warmed to body temperature and above. As you raise a spoonful of warm apple pie to your mouth, your nose immediately begins to detect the volatile compounds that contribute to the unique flavour of the pie. What's more, when you start to chew the pie, those aroma molecules waft down your throat and back up into your nose. The hexyl acetate from the food is registered as an apple flavour, acetoin and diacetyl give us butter and cinnamaldehyde provides cinnamon. Underlying all of this is the sweetness of the sugar and the tart, sour and acidic flavour of ascorbic acid from the apple. Combined, the tastes from the tongue and the aroma molecules detected by the nose add up to apple pie.

What is key here, though, is that if you were to take away the nose, all that is left is sweet and sour. Without your nose functioning, your ability to taste is massively diminished. It's why you can't taste anything when you have a cold. Your nose is bunged up and the flavour molecules can't reach the specialized cells responsible for smelling. This is where we get back to the Maillard reaction, as this produces a whole range of delicious aroma molecules.

While it is true that back in 1912 Monsieur Louis Maillard published the first description of the basic chemical reaction, it was not until much later in 1953 that the full reaction scheme was worked out by John Hodge over in Illinois in the United Sates. The reaction begins when the temperature hits about 140°C (284°F) and a sugar molecule reacts with one of the building blocks of protein, an amino acid. What is crucial is that neither of these ingredients for the reaction need to be free-floating molecules. Sugar molecules in our diet are more commonly joined up in pairs or long chains. Sucrose, or table sugar, is made of the two sugars, glucose and fructose, linked together, while all starch in pasta, potatoes or rice is made of glucose linked into very long chains. In a similar way, proteins are made up of huge long chains of hundreds of amino acids linked together. The Maillard reaction can begin when any sugar at the end of a long chain meets and reacts with an amino acid at the end of its chain. The result of this meeting of sugar and amino acid is a new chemical that spontaneously rearranges itself to form what is known as an Amadori compound. This is where it all gets tricky.

What happens next depends on the exact nature of both the amino acid and the sugar involved in the initial reaction. There are at least six really common sugars that can take part in a Maillard reaction and over twenty different amino acids. On top of that, the reaction that takes place depends on the surrounding acidity and the precise temperature. All of the common products of the Maillard reaction are five- or six-atom, ring-shaped molecules made up predominantly of carbon with possibly some oxygen, nitrogen or sulphur. They have exotic names like pyrazines, furanones, oxazoles and thiophenes, and all of them create interesting aromas. The range of molecules associated with the Maillard reaction gives us nutty, or meaty, or roasted, or caramel flavours. It also yields the brown colour we associate with cooked food.

The Maillard reaction is crucial for so much of cooking. Take, for example, a piece of steak. If you were to cook the meat at low temperature, perhaps in a sous vide cooker (*see* page 35), it would be tender and moist and taste a bit meaty and beefy but not of much more. On the other hand, take the same cut of beef and cook it in a hot pan well above the critical Maillard reaction temperature and the brown crust you form will impart a huge array of flavour molecules and a big boost to the taste. You may wonder how it is that meat, which we are told is made of protein, can undergo a Maillard reaction. Where does the sugar come from? Well, energy within any animal is transported in the blood in the form of glucose, and it is also stockpiled in muscles in long chains known as glycogen. A piece of steak will be packed with more than enough sugar in the form of glucose and glycogen. It's not just meat where a Maillard reaction is flavourful. Flour-based products undergo

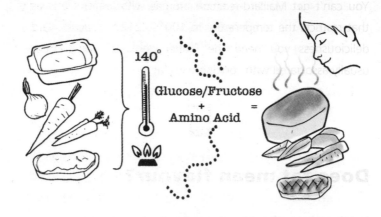

the reaction when they are baked. The browning of a loaf of bread or a bagel is Maillard produced, as are the rich nutty tastes. If vegetables are oven baked or fried you will hit the crucial temperature and create Maillard flavours. So, when you fry onions or roast parsnips the sweet, nutty and caramel flavours are all down to this reaction.

One of the key things to remember about the Maillard reaction is that it only really starts to happen at temperatures above 100°C (212°F). You will get a little bit of reaction at about 120°C (248°F) but it doesn't really kick in until you hit 140°C (284°F). Which means that anything steamed or boiled will not produce these rich aromas. If you are making a stew, for example, and just chuck all the ingredients, the onions, the meat and so on, into the pot with the liquid, it may seem a bit flavourless. Which is why we are told by chefs to brown our meat and onions first in a frying pan at higher temperature.

You can't get Maillard-reaction aromas with water alone as that restricts the temperature to 100°C (212°F). For Maillard deliciousness you need the higher cooking temperatures usually associated with cooking with fat.

Does fat mean flavour?

Here's a phrase that I hear declared by many of the chefs on television: fat means flavour. It is usually exclaimed as they plop a huge pat of butter into a frying pan. They are doing it, they claim, because it makes the dish wonderfully aromatic and flavourful. You get the impression this is almost a way to alleviate their guilt at adding so much fat to a recipe. They know that it is probably not good for you to eat so much fat, but think of the flavour.

You will also find this idea in the cookery books on your shelf: for food to have flavour it must have fat. The idea seems to crop up especially when discussing what qualities make for the best and tastiest meat. I've even seen discussions about how the best meat must have fat and that lean meat is a complete waste of time. It is not even just an anecdotal idea. The US Department of Agriculture has the task of grading the quality of beef produced on different farms. One of the key things seen as desirable in the meat is what is known as marbling.

This is where the lean portions of the beef are laced with thin veins of fat deposits. But does this really improve the taste?

Let's start by nailing down what we mean by fat. First, I should clear up the difference between fats and oils. The only difference is what form they are in at a room temperature of 21°C (70°F). Fats are solid at this temperature while oils are liquid. So, butter is a fat and so is coconut oil, despite its common name. However, what you get from sunflower seeds is definitely an oil.

The chemistry of fats is fairly straightforward: you typically have three chains of about eighteen carbons joined together at one end. The long chains are known as fatty acids and come in a variety of different forms and it is this that gives different fats their characteristics, like what temperature they melt at. It is also within the fatty acids that you find the saturated and unsaturated parts of the fat. The linking portion of the fat is a chemical called glycerol or what used to be called glycerine. You can buy little bottles of glycerol in supermarkets to add to your cake icing to stop it from hardening.

There are, of course, a few more subtleties and variations when it comes to fat and one that is worth mentioning is the fat found in our cell membranes. All of the cells that make up your body, or the body of any living thing, are surrounded by an incredibly thin, wobbly barrier known as the cell membrane. This is constructed from two layers of fat molecules called phospholipids. These are slightly different to the usual fat molecules as they only have two long chains of carbons and in place of the third chain there is a phosphate group that has lots of positive charges. This makes the phospholipids into amphiphiles (see page 88) and they will spontaneously arrange themselves into membranes with all the carbons on the inside of the membrane and the phosphate bits poking out on both sides. We will come back to phospholipids later.

Armed with some basic fat biochemistry, can we answer the question of whether fat means flavour? One answer to this came in 2015 from scientists at Purdue University in the United States. Cordelia Running and Richard Mattes showed that we appear to have specific taste receptors for fatty acids on our tongues. Test subjects, using the taste on their tongues alone, could differentiate solutions containing just fatty acids from other solutions of things like salt or sugar. The new taste was dubbed oleogustus. So, if fatty acid or oleogustus is a basic taste on our tongue, what does it taste like? Well, this is tricky as how do you describe a taste? How would you explain what salt tastes like without actually referring to salt? Clearly oleogustus tastes of fat, but it is also not a good taste and high concentrations of fatty acid were described as awful by the test subjects. The working theory is that our ability to detect fatty acids on our tongues is some kind of early warning system to alert us to potentially rancid food.

So, while fat is a distinct taste, when a chef tells you that fat is flavour, they are not referring to our ability to detect it on our tongue. If fat is not a pleasant taste, can it contribute to the other aspects of flavour that are the aroma molecules wafting up your nose (see page 101)? The answer to this is yes, but not in the way you might have guessed. There has been a lot of work done on how fat contributes to flavour in meat and my favourite example, which underpins much of the later science, was work done back in 1983 at the now defunct Meat Research Institute in Bristol. Food scientist Don Mottram measured the smell of cooked beef with and without fats. Samples of meat had their fat chemically extracted, were then cooked and the aromas of the beef analyzed using a technique called gas chromatography. While this sort of analysis does not tell you how a human nose perceives the smell, it can identify how smells change. When the beef had all the visible fat removed, the profile of aromas produced when it was cooked did not change. Which means that the fat was not contributing to the smell. However, if you take out the phospholipids, the special fat molecules that make up membranes, the smell did change. So, in this completely artificial set-up, the ordinary visible fat in meat made no difference to the taste. The only flavour contribution from fat came from the hidden phospholipids that are found in all meat, even the lean stuff.

If this is the case, what does fat contribute to our sensation of eating? The fat itself does not have much in the way of a flavour profile, but there are flavour molecules, produced by things like the Maillard reaction (see page 100), that dissolve in fats. Since fats tend to hang around in your mouth even after the bite of food you were chewing has been swallowed, you have more chance to detect them. This is particularly noticeable with smoked foods like bacon. The smoke flavour molecules dissolve in the fat and cling to your mouth leaving a persistent smoky aroma. However, mostly the contribution of fat is in what is known as mouth feel. Fats give a food a silky or creamy texture in your mouth and lubricate the action of chewing. Foods with no fat that require significant chewing, like a piece of cooked chicken breast, can be dry and unpalatable in the mouth, whereas meat that has more intrinsic fat, like chicken leg, do not have the same dry mouth feel.

While fat itself is not flavoursome, it can help enhance the availability of flavours and it definitely improves the mouth feel of a food. That said, you don't need much fat to achieve all of these things. So, when that TV chef is loading up the dish with butter, you don't necessarily need to follow suit.

The queen of kitchen chemistry

If the Maillard reaction is the king of flavour then its queen must surely be caramelization. Take some ordinary sugar, pop it in a pan and heat it up. Quickly the sugar begins to melt, bubble and, as it turns colour, first becoming a golden brown, the delicious smell of caramel wafts up from the pan. The process of caramelization takes sugar alone and turns it from plain old sweetness to an array of mouth-watering and tempting aromas. It is also crucially the other way to brown our food, a colour change we associate with improved flavour. Like the Maillard reaction (see page 100), it is a complicated process, but at least you start with just one ingredient.

Ordinary white table sugar is composed of pure crystals of sucrose extracted from either sugar cane or sugar beet. Each sucrose molecule is itself made up of two individual sugar molecules stuck together, one glucose and one fructose (see page 102). The first step in making caramel is to break down the double sucrose sugar into the individual sugars. Sucrose is too stable to undergo caramelization directly, but the single glucose and fructose are much more reactive. This breakdown happens spontaneously when the sugar crystals in your pan hit 170°C (338°F) and the reaction requires the addition of water. Which is why many caramel recipes specify the addition of a spot of water to get this first stage to happen. The result of this step alone, by the way, is known as inverted sugar and is a mixture often used in confectionary. So far so good, but I'm afraid that is the only simple part of the chemistry of caramelization.

At this point, the glucose and fructose can react in a few different ways. The most obvious is for them to begin to fall apart. The heat energy causes the individual sugar molecules to break apart and create smaller fragments. Just like with the Maillard reaction, the range of possible products is huge and the flavours they impart equally broad. The aroma molecules created can smell fruity, flowery, buttery or milky and some impart a distinctive roasted note. Keep going with the reaction and the delicate flavour molecules themselves break down, resulting in sour- or bitter-tasting compounds. As the reaction progresses and the caramelization goes further and further, the sour- and bitter-tasting products begin to accumulate. As the sugar breaks down, the resulting caramel gets more flavourful, but less and less sweet. If you let caramelization go too far, all the sugar is used up, and the result is predominantly bitter and not particularly sweet.

Exactly what aromas you end up with depends on a few factors. One of the most critical is the exact proportion of sugars present. In theory, if you have pure sucrose as a starting material you will end up with an even mix of fructose and glucose. But if you are cooking fruit, say you are frying some slices of apple, there will be much more fructose present and fructose breaks down at a much lower temperature. The caramelization of fructose takes place at 110°C (230°F) while glucose hangs in until 160°C (320°F). Given this, is it much easier to overdo the caramelization of fructose and take it too far into the bitter end of the process. The presence of other chemicals will change the aromas produced. Acid conditions speed up the process, as will any water. On top of that, if there is any protein present, the Maillard reaction kicks in at 140°C (284°F) and will start producing its own array of flavours.

None of this, however, explains the colour changes associated with caramelization. All of the broken down products from the sugars are colourless and volatile. For the distinctive colour of caramel, we need to start by sticking the fructose and glucose back together. This may seem crazy, but the way the two single sugars glue themselves up is different to the way sucrose is made. Rather than just being joined at one point, the two sugar molecules create two bonds and a ring structure. What's more, it does not need to be the neat joining found in sucrose with one fructose holding hands with one glucose. Any two sugars can hook up, glucose to itself or fructose together. This peculiar new molecule can react with itself to give rise to one of three different, bigger molecules, each with a different name, known as caramelan, caramelen and caramelin. The caramelan/en/in will then spontaneously reorganize itself and other molecules of the same sort to produce tiny little specks of solid, brown-coloured material. It is these that give caramel its colour. The caramelan particles are about half of one-thousandth of a millimetre (0.02 thousandths of an inch), the caramelen twice that size and the slightly darker brown caramelin comes in at a whopping five-thousandths of a millimetre (0.2 thousandths of an inch). These tiny motes of bound-up sugar give the caramel its distinctive colour, and the longer the heat is applied, the more the particles accumulate and the darker the colour.

Taken together, this suite of chemical reactions takes a simple white sugar and turns it into a complex and deep array of aroma and taste. But it is not just plain white sugar that caramelizes – any sugar will undergo the same reactions. Fruits will readily caramelize as they are chockful of fructose. The same happens with vegetables that contain a decent amount of sugar, such as onions for example. Biscuits cooked at a high enough temperature of 170°C (338°F) will turn brown with the products of caramelized sugar. Drop the temperature below that and, while you get the Maillard reaction, there will be much less browning and no caramelization.

If you take any cooked food item and look at where the flavour comes from, some of it will be inherent, but for nearly all cooked items most of the aromas can be traced back to these two fundamental pillars of cookery: caramelization and the Maillard reaction.

Crystalline complexity of chocolate

There are other, more specific sources of deliciousness that have nothing to do with Maillard and caramel. According to the latest genetic analysis, the tropical tree *Theobroma cacao* is native to the area around the Peruvian city of Iquitos, sat on the banks of the Amazon River. These days, though, it can be found growing in the tropics across the world. It's a huge cash crop and the leading producers are now in West Africa, the Ivory Coast and Ghana being the major growers. The tree produces orange fruit up to 30 centimetres (a foot) long that grow directly from the trunk of the tree. Once ripe, the pods are harvested and the internal white pulp containing the large seeds is allowed to ferment for a few days. After being suitably stewed in their own juices, the seeds are picked out and dried. At this point the seeds are about 2 centimetres (about ¾ inch) long and a peculiar dark purple colour. Then comes roasting and, after the shell is cracked off, the remaining nibs are warmed to about 40°C (104°F) and ground under heavy rollers for a couple of hours. The result is a thick, unctuous, dark brown gloop that is known in the trade as liquor. In case you had not twigged yet, this is the starting point for the production of chocolate.

To make a bar of chocolate from the liquor, in theory, all you need do is add sugar and allow it to cool in a mould. However, the resulting bar of confection is going to taste rather peculiar. It will have a gritty texture and the flavours of the chocolate, the sweetness of the sugar and the bitterness of the cocoa will seem almost separate in your mouth. The bar will also not have the crisp snap we normally associate with chocolate and it will seem a bit greasy and almost chewy.

The sugar and chocolate liquor mixture needs to go through a process known as conching, invented by a certain Swiss gentleman by the name of Rodolphe Lindt in 1879. It's very simple really: you put the sugar and liquor mixture back into the heated grinding machine and leave it running for much, much longer. How long depends on how smooth you like your chocolate but it can easily be twelve hours and up to seventy-eight hours for superfine chocolate bars. The heavy rollers take the size of the sugar and cocoa particles down to about 0.02 millimetres (less than one-thousandth of an inch). Once the particles get down to this tiny size, the gritty mouth feel of the chocolate disappears. It was this discovery by Mr Lindt that made chocolate a popular sweet treat. Before conching, the grittiness of chocolate bars stopped them from becoming popular and mostly people consumed chocolate as a hot drink.

So, now we have superfine particles of sugar and super-fine and flavourful particles of cocoa, both of which are suspended in cocoa butter, which is just the fat from the cocoa bean. And yet, despite the conching, our chocolate bar still does not have the satisfying snap and gloss. To get this, we need to explore the many crystalline forms of cocoa butter.

When you grind up roasted cocoa beans, a little over half of what you get is a fat known as cocoa butter. It's often extracted from the chocolate liquor by pressing it through a fine filter. The resulting silky-smooth fat gets used in all manner of cosmetic products like face creams and soaps, but mostly it gets fed back into the chocolate industry. White chocolate, for example, is just cocoa butter mixed with sugar and dried milk powder. As an aside, what you are left with once the cocoa butter has been filtered out of the chocolate liquor is a hard-pressed cake of cocoa. This cake is then pulverized to create cocoa powder and sold for use in baking and to make hot chocolate drinks.

Cocoa butter is remarkable for several reasons. It has a very high melting point for a vegetable-derived fat and does not turn to a liquid until it hits a temperature of 34°C (93°F), which is really important as this is just below body temperature of 37°C (99°F). Since it is the cocoa butter that holds a block of chocolate together, when you pop a chunk in your mouth the fat melts and all the flavours flood across your tongue. But there is another thing that makes cocoa butter a bit special. Since it has a very regular structure it can arrange itself into neat crystalline structures. This may seem a bit strange as when you think of crystals you usually think of shiny gems, but any molecule that can pack itself into a regular three-dimensional array is forming a crystal. What makes cocoa butter unusual is that it can form itself into not one type of crystal, but six. The secret to putting the gloss and the snap into a chocolate bar is making sure the fat is in the right crystal form. Crystal types one to four are no use to us as they give a crumbly texture to the chocolate. On top of that the melting point of these fat crystals is much lower, ranging from 17°C (63°F) to 28°C (82°F). If you have a chocolate bar with too much of type-one to -four crystals, it will start softening at room temperature and break up more like cheese than chocolate. The type-five crystal structure, on the other hand, is just right. When cocoa butter fat forms type-five crystals, the fat molecules pack closer together than in the first four types. If this is the dominant form, then the chocolate will have a crisp snap to it and a glossy sheen, so long as the crystals are small. The final type of crystalline structure, type six, also gives a good snap and gloss to the chocolate, but it takes so long to form that it is impractical and rarely encountered.

How then do you make a bar of chocolate and ensure that most of the fat forms into type-five crystals? If you mix up a batch of cocoa liquor with sugar and then allow it to cool straight from the conching machine, the cocoa butter fat will spontaneously start forming all of the first five crystal structures. To stop this and force type-five crystals to dominate, a process known as tempering is used. There is a bunch of different ways this can be done, using pre-tempered chocolate or a marble slab for example, but so long as you have a trusty digital temperature probe to hand (*see* page 35) there is one way that takes out all the guesswork.

Place your chocolate in a metal bowl on a pan on hot water, then start by raising the temperature of the chocolate to 50°C (122°F). At this temperature, all of the six forms of cocoa-butter fat crystal will have melted. Take the bowl off the heat, watch the temperature probe and start to stir. The act of stirring makes sure your chocolate is an even temperature and also keeps any cocoa-butter fat crystals small. The first key point you are watching for is when the temperature hits 34°C (93°F). This is the temperature at which type-five crystals melt. So, once you get below this temperature you will start to form type-five crystals of the cocoa fat, but none of the other crystal types. Keep stirring and allowing the mixture to cool, until you hit 28°C (82°F). Don't let it go below this. If you do, you are in the temperature range where the undesirable types of fat crystal begin to form. Your chocolate should have thickened considerably as the fat begins to form tiny type-five crystals. Now put the chocolate back on the heat and raise the temperature to 32 °C (90 °F). This is just below the 34 °C (93 °F) melting point of type-five crystals. So, your precious crystals won't melt and will slowly keep growing, setting the chocolate, but since you are only just below the melting point, the molten and now tempered chocolate will hold at this temperature for hours. When you are ready to set the chocolate, pour it into a mould or over a cake or whatever you have planned for it. Then rapidly cool it down to 15 °C (59 °F). As the molten chocolate begins to cool the fat starts to crystallize. Since, by the action of tempering, you have ensured the mixture is full of billions upon billions of tiny type-five crystals, these will grow and predominate. The final product will have a much closer-packed crystalline structure and consequently will be harder, and have a satisfying snap and a glossy sheen.

This is not, however, all the chemistry lurking within the dark and delicious bar of chocolate. While the crystals of fat explain the snap and gloss, they do not cover why we crave it so much. Well, by we, I possibly mean me. Is it possible to be a chocoholic? Is there a chemical reason for the desire?

The chemical often hailed as being responsible for this is the somewhat confusingly named theobromine. The confusion comes because it contains no bromine, just carbon, nitrogen, oxygen and hydrogen. Its name is derived from the Latin for the cocoa plant, *Theobroma cacao*, which in turn gets its name from the Greek for god (*theo*) and food (*broma*). Chocolate is literally the food of the gods. From a pharmacological

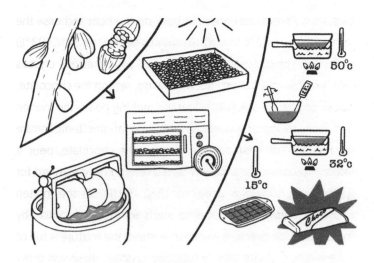

Chocolate from pod to bar.

perspective, theobromine acts a bit like caffeine. Consequently, it increases your heart rate, but also makes your blood vessels relax and thus reduces blood pressure. Like caffeine, it also causes an increase in urination and can prevent you from getting to sleep. It's hard to imagine, but it is possible to overdose on chocolate and get too much theobromine in your system, at which point you will suffer nausea and vomiting. Humans have it easy though, as we are very good at breaking down theobromine. Some animals cannot break it down. Dogs, for example, can easily suffer from theobromine poisoning. The drug will stay within their blood for hours and as little as 60 grams of chocolate per kilogram (about 1 oz per lb) of the dog's body mass can be lethal. Interestingly, theobromine is dangerous to cats, too, but poisoning is rarely seen as, unlike dogs, cats cannot taste sweetness and have little incentive to eat chocolate in the first place. While there is no doubt that if you eat a big block of dark chocolate, rich in theobromine, there will be an effect on your body, there is nothing about the substance that makes it addictive.

For that, there are two other possible suspects we should consider. The first is an amino acid, a building block of proteins, called tryptophan that is used by our bodies to manufacture another drug known as serotonin. This is secreted in our brains to give a feeling of well-being and happiness. This sounds like a good candidate for the addictive nature of chocolate. Unfortunately, the most recent studies show that while an increase of tryptophan in your brain does give rise to an increase in serotonin, this does not happen if you just eat the stuff. Tryptophan is sent to your brain through a special amino-acid transporter. The problem is that when you eat chocolate you ingest a whole pile of different amino acids and these effectively block up the amino-acid transporter. The end result is that chocolate does not increase the levels of tryptophan in the brain.

The only other suspect to explain chocoholism is phenethylamine, a naturally occurring chemical found in many plants that is probably being produced by the plant to act as an antimicrobial agent. On its own, phenethylamine is a psychoactive drug and within the drug industry it has been used as a starting point for the manufacture of a whole range of compounds, including psychedelics, anti-depressants, stimulants, anti-anxiety drugs and, somewhat randomly, a range of decongestants. But once again, the chemicals within chocolate are foiled by our own bodies. All of the effects of phenethylamine and its derived drugs come when injected directly into the blood. If you eat it, it is rapidly broken down in your intestines and significant amounts never reach your brain no matter how much chocolate you scoff.

Why then do we crave chocolate? It seems from most recent studies that we do actually crave the stuff. This is not just greed: chocolate does have a distinct addictive quality. However, no matter what the newspapers report, there is no pharmacological effect at play here. At least none that is different to why we crave sugars and fats generally. The current idea is that it is the hedonistic and psychological appeal of chocolate that is the cause. We crave the sensation of the sugar and fat, the smell and the very idea that we are giving ourselves a treat. I am afraid that for those of us who struggle to resist chocolate, we really do have no one else to blame but ourselves.

What's wrong with sprouts?

If the science behind a love of chocolate is lacking in hard chemical facts, how about a dislike of Brussels sprouts? Of all the vegetables that can appear on a dinner table, the Brussels sprout has to be the most contentious, especially among children. They are the subject of innumerable jokes, not least those contained within Christmas crackers. Clearly, they are not universally hated as, come the festive season, my supermarket has great stacks of them. What splits the population on the subject of sprouts?

The secret to the sprout appears to be a protein called 'taste receptor 2 member 38' that is found sitting in the taste buds in your tongue. This receptor is responsible for detecting some bitter-tasting compounds in your food and reporting their presence to your brain. The problem is that the gene in your DNA that encodes this protein has a few variants. The gene has the unmemorable title TAS2R38 and the two variants of this gene are known simply as taster and non-taster. Now, the DNA that is present in every single cell in your body is there as two copies: one you inherited from your father and the other from your mother. When your taste buds start making the protein encoded by TAS2R38 they will use both copies as their template. If you have even one of the taster genes in your DNA then you get a functional taste receptor 2 member 38 protein. If both your mother's and your father's copy of TAS2R38 that ended up in your DNA are the non-taster form then the protein you get doesn't work and your ability to detect bitterness is reduced. Since the only way you end up overall as a non-taster is to have a double hit of the non-taster variant of the TAS2R38 gene, this is usually much less common in the population. Exactly how many people are overall non-tasters depends on where you come from; in

Europe it averages about 28 per cent, in China it's only 14 per cent while among Australian Aborigines a whopping 50 per cent are non-tasters. The way this is tested is not by feeding people sprouts, but with a more precise chemical test. There are a couple of synthetic chemicals that only tasters can taste: phenylthiocarbamide (PTC) and the more commonly used propylthiouracil (PROP). These days you can buy packets of little paper strips impregnated with PROP to test out if you are a taster or not, and it is usually done in school biology classes as a simple population genetics experiment.

The working hypothesis then goes that there is an unpleasant bitter chemical in sprouts that only some of us can taste because of our genetics. People with non-taster genetics don't have the problem and enjoy eating sprouts. The most likely candidate for the bitter chemical is known as glucoraphanin, or it could be progoitrin, or there is an outside chance of it being sinigrin. There is also the possibility that it is all three of these or none of them and something entirely unknown to science. Right now, as I write, the jury is out on this one. However, we do know that whatever it is, it falls within a group of chemicals known as glucosinolates. The general structure of these is as follows: they contain a glucose molecule, connected through a sulphur atom to one of the building blocks of proteins called amino acids, which in turn is connected to another sulphur atom. Which amino acid you have is what makes the difference between glucoraphanin, progoitrin or sinigrin. All members of the brassica plant family, which includes the Brussels sprout, produce loads of different glucosinolates in their leaves, and because of the shared molecular structure they react in similar ways.

When you chew on a Brussels sprout, you break open the cells of the sprout and all the different compartments within the cells. The glucosinolates mix for the first time with enzymes that start to chop up their complex chemical structure. Irrespective of the differences, you end up with one crucial compound, an isothiocyanate. This has a sulphur atom, connected to a carbon, connected to a nitrogen, connected to a variable tail bit. It is this chemical that the PROP and PTC chemicals mimic in taste tests. The isothiocyanate is detected in our mouths by the product of the taster version of the TAS2R38 gene, and we sense this as bitterness. If you are a non-taster with the other version of the TAS2R38 gene, you taste no bitterness.

So, if you are taster you don't like sprouts and if you are a non-taster you do? Well, not quite; as with anything to do with humans and biology, it is more complex than that. The reason plants go to the effort of making all these chemicals is fairly straightforward. What you have here is a defence against herbivory and an attempt by the plant to stop animals from eating it. However, the perversity of humans means we often seek that which harms us, or at least tastes bitter. Young children have an innate dislike of anything bitter, for good reason as it often indicates potentially toxic chemicals are present. However, as we age we get over this dislike and start to seek out bitter flavours: as an example, the dominant taste of a cup of coffee is bitterness. Many people who like Brussels sprouts enjoy their flavour even though they are tasters, myself included. For decades, I was not a fan of sprouts and dodged them at the Christmas dinner table. Until, that is, I discovered that what I didn't like was overcooked sprouts. I had a Brussels sprout epiphany. It turns out I really like sprouts, just not boiled ones. If you trim your sprouts, quarter them and then steam for no more than five minutes, then maybe toss them in melted butter and sprinkle with toasted almonds, they are delicious. At least I think so. They have a nutty and, yes, slightly bitter taste in my mouth, which I really like.

So, what is it then that I dislike about sprouts cooked for longer? Firstly, the texture begins to soften and I've never been a fan of mushy vegetables. If you cook sprouts whole, the spherical shape means that it is inevitable that by the time the middle is just cooked the outside is overcooked. Which is why I quarter my sprouts. But there is another bit of chemistry that makes me and many others turn their noses up at sprouts. All those glucosinolates packed into the leaves of the sprout are fragile molecules and, if you heat them too much or for too long, they fall apart spontaneously. One of the results of this disintegration is a gas we commonly associate with the sulphurous smell of hard-boiled eggs. It's called sulphur dioxide and our noses are extremely sensitive to it. If this is a smell you dislike, then I would guess that you're not a fan of egg-mayonnaise sandwiches and any overcooked cabbage relatives, Brussels sprouts included.

It turns out that sprout avoidance is a complicated bit of science. However, underlying all of this, and I suspect one of the most crucial issues, is how the sprout is cooked and presented. My guess is that the sprout dodging is more to do with cookery and less to do with science.

A stimulating brew

Caffeine is the most widely used psychoactive drug on the planet. Consumption of caffeine changes your mental state by preventing drowsiness and increasing alertness. It also has more general bodily effects, boosting athletic performance and improving coordination. On the negative side, it causes vasoconstriction (your blood pressure goes up), increases gastrointestinal motility (it makes you poo), heightens gastric acid secretion (it can give you heartburn) and, in large doses, it is a diuretic (it makes you wee and you become dehydrated). While it is not technically addictive according to the accepted psychological definitions of the term, it does cause mild physical dependence. There are withdrawal symptoms and some neurologists argue that this means its use and abuse can lead to a psychological disorder. Which all makes it sound rather hazardous, except that it is estimated by the FDA (the US Food and Drug Administration) that 90 per cent of the North

American adult population consumes some form of caffeine each day. Globally, in developed countries, the picture is similar with a few local differences in the form we prefer to consume it. Most caffeine comes from coffee, although here in the UK we consume an inordinate number of cups of tea and many of the British get their caffeine this way. Given its popularity and ubiquitous penetration of our societies, it's no wonder that caffeine-containing foods are held in high esteem by many of us.

The coffee bean comes from two different but closely related plant species. There is *Coffea arabica*, which makes up 70 per cent of global production and is considered to make superior coffee, and there is *Coffea canephora* var. *robusta*, which accounts for the other 30 per cent and gives a more brutish but cheap-to-produce coffee drink. Both plant species grow to about 10 m (33 feet) tall, with dark, glossy green leaves and delicate white flowers that produce little clusters of 15 mm (about ½ inch) berries that ripen to a dark red. The *arabica* plant is hard to cultivate but worth it as it produces the highest-regarded coffee with a mild, rich and subtle flavour. The *robusta* species is, not surprisingly, a hardier plant. It can grow in shade or full sun, produces bigger crops that contain more caffeine, needs less fertilizers and suffers from fewer pests, but the coffee it makes does not taste as full-bodied. It's used in cheaper instant coffee brands, but also there is often a bit added to espresso coffee mixes as it gives a good kick to the blend.

Once the berries, or cherries as they are known in the trade, are harvested there are several ways to process them, but the simplest involves leaving them out in the sun for a few days on large sheets to dry out. Once suitably dry, they are fed into a machine that carefully cracks off the dry cherry flesh and leaves behind the green coffee beans, two from each cherry. At this point, if you try to eat one it's a bit like chewing on un-popped popcorn kernels: very hard and essentially tasteless. The flavours are all developed in the roasting process. It is an easy enough process to do – if you can get your hands on green coffee beans you can give it a try in a dry frying pan. The issue is not to overcook the beans, and getting it just right is a subtle science. As the bean heats up and gets to 100°C (212°F), what little water remains vaporizes and the bean puffs up a little bit. The texture is now crumbly and full of tiny holes filled with water vapour. When you hit 140°C (284°F), our old friend the Maillard reaction kicks in (see page 100). The green bean turns brown and flavour molecules are produced. Heat further to 160°C (320°F) and the heat generated by the Maillard reaction itself becomes self-sustaining and drives the temperature rapidly up to near 200°C (392°F). At this point, another gas is produced, carbon dioxide, which pushes the water vapour out and fills all the holes in the crumbly bean.

When the bean is just a light brown colour, the chemistry going on produces lots of acidic compounds, making any coffee you make from these beans taste tart or sour. As the roasting continues, the bean darkens to a medium brown and the acids break down, reducing sourness. At the same time, you get the characteristic flavour molecules of coffee developing along with a little bitterness. But beware, if you take the beans to a darker roast, more bitterness develops and many of the delicious coffee aromas are masked. All of this chemistry, from acidic coffee to over-roasted, takes place over only 30°C from about 190°C to 220°C (54°F from 374°F to 428°F). The whole process of flavour development can happen in just a matter of seconds. Which is why it is tricky to do at home and best left to the experts with equipment a bit more controllable than a frying pan.

One of the good things about this process is all the carbon dioxide in the beans. Since this drives out the water and oxygen from inside the beans, once roasted they keep for a relatively long time at room temperature without spoiling (a couple of weeks) when compared to ground coffee (a few days). It's also this trapped carbon dioxide that produces the distinctive crema froth on top of an espresso. As high-pressure water is forced through the ground espresso coffee, the carbon dioxide gas escapes and forms tiny bubbles held stable in the oils from the ground coffee.

About 1 per cent of the dry weight of a roasted *arabica* coffee bean is caffeine. Beans from *robusta* can have significantly more and up to 4 per cent can be pure caffeine. Drink a cup of black coffee and you will have ingested about one-tenth of a gram of caffeine (three-thousandths of an ounce). Once this gets into your system it has a number of effects, all of which rely on its ability to mimic adenosine, a chemical with which it shares some structural similarities.

Within our bodies adenosine plays an important role as a neurotransmitter, used as part of the system whereby one nerve cell transmits a message to another. In a normal situation, when an electrical signal travelling along a nerve cell gets to the end of the nerve, it causes a little bit of neurotransmitter, such as adenosine, to be released from the nerve tip. The adenosine then sticks to a special receptor on a neighbouring nerve cell tip and starts a new electrical signal in this neighbouring nerve. And so the message is passed on from one nerve to another. When you wake up in the morning, your brain has very little adenosine inside it. As the day wears on though, adenosine is produced in nerve cells and begins to accumulate. When adenosine is released as a neurotransmitter, the effect it has is to signal to your brain that it is weary and tired. Elsewhere in the body, it reduces your heart rate and makes your blood vessels expand, which drops your blood pressure. It's basically an all-round sleepy chemical that slows your body down.

The really cunning thing that caffeine does is to imitate adenosine and bind to all its receptors on your nerves but crucially, and unlike adenosine, when it binds to the receptor nothing happens. No new signal starts in the nerve cell when caffeine binds to it. Which means that if you drink lots of coffee, the caffeine gums up all the receptors and the normal adenosine sleepy signal is blocked. The result is that you stay wakeful and alert. It also blocks the way adenosine slows your heart and drops your blood pressure, which is why caffeine gives you a feeling of being energized. While caffeine has the same result as a stimulant drug, it does this by turning off your innate slow-down system. Consequently, your body's natural stimulants like dopamine and adrenaline keep you active and awake.

All of which explains what happens when you drink a cup of coffee and get a caffeine buzz. But why do so many of us crave it? Since caffeine is not a stimulant itself, you don't get addicted in the same way you would to drugs like cocaine or amphetamines. But you may well grow to enjoy the feeling of being awake at a time when your body is tired and trying to point you towards bed. If this is the case, you will associate coffee, or a caffeine source of your choice, with those positive feelings and a psychological craving is set up.

However, there are some physical changes in people who drink coffee. Even a single cup of coffee a day will cause your body to start making more adenosine receptors. The more receptors you have, the more caffeine you need to block them and the less of a buzz a single cup of coffee will give you. Over time we become accustomed to caffeine and its pick-me-up effects are diminished. What's worse is that if you now stop drinking your daily coffee, you have an over-abundance of adenosine receptors with nothing to block them. You are now going to be more susceptible to the sleep-inducing effects of adenosine and feel more tired. On top of this, the excess adenosine activity gives you withdrawal symptoms. Most commonly, people will suffer from headaches. All the extra adenosine receptors are free from caffeine and act to relax your blood vessels, allowing the blood to flow slower and that gives you a slight swelling of the surrounding tissue. When this happens in your brain, this tiny swelling gives you a headache. Of course, once you have had a cup of coffee, the adenosine receptors are blocked, the blood vessels constrict, the swelling goes down and your headache passes. The other commonly reported symptom of caffeine withdrawal is being irritable. Although I'm always somewhat cranky when I have any sort of headache, so I'm not sure how you can tell what causes the crankiness: too many adenosine receptors or the caffeine-withdrawal headache.

Like any drug we use, whether it is for recreational purposes or as a pharmaceutical to treat something pathological, the poison is in the dose. Caffeine is poisonous if you consume enough of it. What constitutes enough depends on a few things like how much you weigh and how quickly you consume it. Theoretically, 100 cups of coffee have enough caffeine to kill a grown man. But since caffeine is cleared from your body in about four hours, you would have to drink all that coffee at a blistering pace. More worryingly, you can easily buy pure caffeine powder as a sports performance enhancer. A heaped tablespoonful of this would constitute an overdose.

So, yes, we crave coffee and the caffeine it contains, mostly because we like the feeling of being awake, alert and full of energy it gives us. It is not a physical addiction but once you are on the caffeine treadmill, getting off can prove to be, quite literally, a headache.

Sharing Our Food With Bugs

The five-second rule

The five-second rule goes like this: if you drop some food on the floor, it is OK to pick it up again and eat it so long as it was on the floor for less than five seconds. Occasionally you see a stricter version of this: the three-second rule. But is there anything even remotely scientific in either of these nuggets of folk wisdom?

Clearly you don't want to eat any bits of adhering grit or strands of hair, but let us assume that you brush or blow away any visible detritus picked up from the floor. Is the food safe to eat? There is a very simple answer to this question: no, the food is unsafe for consumption once it touches the floor. But when examining this answer, we first need to think about what we mean by safe. Because if this is about safety then it is really about risk, and if it is about risk then it's time for some risk assessment.

If you put on your health-and-safety hat, then the first thing you do when risk assessing is to identify the potential hazard or, in other words, what it is that could go wrong. In this case, assuming the food is free from inedible grit, then the danger is that you may be about to ingest harmful bugs that cause an upset stomach, probably diarrhoea, stomach cramps and possibly a fever. The most likely cause of this is bacteria called *Campylobacter jejuni*, which most people haven't heard of because the food poisoning it causes is normally pretty minor. However, about three-quarters of all upset stomachs are caused by these bacteria. The ones that make up the other quarter of cases are the bacteria you have to watch out for. Food poisoning caused by salmonella and *Escherichia coli* (or *E. coli* for short) can be really nasty. In the UK, about 2,500 people end up hospitalized each year with salmonella. But the one you really don't want to pick up is *E. coli*. Normally this is a harmless bacterium that we have living happily, and harmlessly, in our guts (*see* page 137). But there are a few strains of this bug that have developed a particularly nasty ability to produce Shiga toxin. Named after an early twentieth-century Japanese microbiologist who first described it, Shiga toxin has some extremely unpleasant effects on our bodies. I won't go into all the gruesome details, but the end result is often hospitalization and it can lead to kidney failure and even death in extreme cases. The most well-known strain of bacteria that can cause this is known as *E. coli* O157:H7 and it's been responsible for some high-profile outbreaks all over the world. The somewhat scary thing is that you don't need that many of these bacteria to give you a dose of food poisoning.

When you drop food on the floor, the hazard is that you then ingest something like *E. coli* O157:H7 and die. The second part of any risk assessment is to look at how likely it is that the hazard is actually going to happen, and this is where the science comes in. There have been a number of studies on the topic of the five-second rule, going back to 2003 when the first experiments were done by Jillian Clarke, a sixteen-year-old summer intern at the University of Illinois in the US. She took small square tiles, inoculated them with a harmless variety of *E. coli* and then dropped either a gummy bear or a cookie onto the tile. After five seconds the food was removed and in every case was found now to be contaminated with the test bacteria. For her efforts, she won the 2004 Nobel Prize for Public Health. Well, to be honest it was an Ig Nobel Prize, awarded for making people laugh and then think, and she shared the prize stage with studies on the physics of hula hoops, a man who patented the comb-over hairstyle and the group that discovered that herring fish communicate by farting. What was important, though, was that she took what many would assume to be either a trivial or silly subject and applied rigorous science to it.

After her initial study, other researchers took up the baton and have expanded on the subject, testing different foods dropped on to a range of surfaces. This is also not fringe science; the latest work was accepted for publication at the end of 2016 by the highest-rated scientific journal in the field of microbiology. In this study, the scientists, Robyn Miranda and Donald Schaffner at Rutgers, the State University of New Jersey, used a range of foods including bread, buttered bread, slices of watermelon and, once again, gummy bears. It may seem like a strange mix of food to drop, but it covers a range of foods that are wet, oily, dry and, well the gummy bears are a bit of mystery to me, to be honest. They dropped the food on to carefully inoculated sections of steel, ceramic tile, wood and carpet and then looked at how many bacteria transferred after a second, five seconds, thirty seconds and five minutes. What they found backed up Jillian Clarke's work: the food was contaminated the moment it made contact, or at least within a second. Which means that the five-second rule is rubbish, but they also found that the longer you left something the more bacteria transferred on to the food. What comes as a surprise is just how few bacteria there are on a surface that is dry and has been dry for a considerable time. Bacteria can survive in dry conditions, but not for long. If a surface has been left dry for a few hours or preferably days, the number of bacteria it may harbour is really small. However, if your surface has a film of water across it, chances are it is teeming with bugs. Similarly, wet food is much better at picking up bacteria. The water is sticky and flows into all the nooks and crannies of the surface on to which the food is dropped.

So, it seems that the odds change if the food you drop is dry and not sticky – maybe a bit of toast – and if the surface you drop it on is also dry and has been dry for many hours, perhaps the floor in your kitchen. In this scenario, the chances are that there are very few bacteria to pick up and not many of them will transfer on to your toast. In which case, you may decide that you can eat it. However, although you have probably minimized the chances of anything untoward happening, you can never reduce them to nothing. And don't forget it only takes a few *E. coli* O157:H7 and you can become seriously ill.

I should also point out that the work by Miranda and Schaffner discovered that the best thing to drop your food on to, for minimal contamination, was carpet. Presumably the food makes very little contact with the actual surface as it sits on the upthrust carpet fibres. If the toast falls on carpet, the odds are in your favour. However, you may end up with fuzzy toast and, if the toast lands butter-side down, then it's just a horrible mess.

The microbiota in us and on us

All of this talk of deadly bacteria that you should never take a risk on may give you the idea that all bacteria are bad for us, but that is not the case. Bacteria are an incredibly important part of the way we digest our food and, recently, it has been suggested that they may even be involved in controlling our appetite for food. It has been common scientific knowledge for many years that there are millions and millions of bacteria living on our bodies and inside our bodies. Initially it was thought that these bacteria were essentially hitching a ride with us and, while they caused us no harm, they were equally of no great benefit. You will see various estimates as to just how many bacteria there are, but the media and popular kids' science books often revel in the statistic that there are ten times as many bacterial cells in a human being as there are human cells. Which implies a great conclusion that we are numerically more bacterial than we are human. The numbers are also really big, which always helps attract attention. The bacterial number usually quoted is 100 trillion cells, that's a one with fourteen zeros after it. The most recent thinking is somewhat less dramatic and the latest estimates done in 2016 by a trio of scientists at the Weizmann Institute of Science in Israel found that the ratio is much less than ten to one and more like one to one. On average, for a standard 1.7-metre tall, 70-kilogram man (that's 5 feet 7 inches and 154 lbs) there were slightly more bacterial cells than human cells, the ratio being 1.3 to 1, which does not make nearly as good a headline.

But what is interesting is that the number of bacteria vary wildly between people; some will have twice as many bacteria as the average and some half the number. The massive downgrading of the numbers is mostly due to better data but also down to an understanding of where bacteria are found. Comprehensive surveys have been carried out and the bacterial populations mapped on a wide range of individuals. Different parts of your body host different bacteria. The bacteria on the skin of your scalp live in a radically different environment to the ones between your toes; consequently the types of bacteria found are also completely different. Belly-button bacteria are worthy of a quick mention. Apparently, the environment in the average belly button, or umbilicus to use its proper name, is such that only a single family of bacteria can survive. It is clearly an unusual environment. So much so that on one of the people tested, researchers found members of an entirely different domain of life, the archaea. To clarify, there are three domains of life: bacteria, eukaryota (which includes plants, fungi and animals) and the archaea. These are similar to bacteria but have unique biochemistry and are normally only found in the most extreme environments. That said, the person with the archaea living in their belly button did claim that they had not washed for several weeks, which may explain the extreme environment.

Taken together, all of the bacteria living on and in our bodies are known as microbiota. So, what are they all doing? The idea that these are just hitchhikers on a human body is no longer believed. Your intestines house the bulk of these bacteria, mostly concentrated in the large intestine as the small intestine and stomach are generally too harsh an environment for many bacteria to thrive. It turns out that the bacteria in your gut play a number of essential roles. Firstly, you have bacteria in your large intestine that are capable of digesting some of the plant fibre and complex carbohydrates that would otherwise be completely indigestible. These bacteria break down the fibre to produce what are known as short-chain fatty acids, or SCFAs to their friends. These in turn can now be absorbed by your body and provide energy and help you gather essential nutrients such as calcium, magnesium and iron. You will have encountered the effect of what happens when this bacterial digestion goes wrong. If you take a course of antibiotics, a common side effect is antibiotic-associated diarrhoea. The drug not only kills off the pathological, bad bacteria, but also the good ones, so that your gut bacteria are essentially wiped out, fibre cannot be turned into SCFAs and water is trapped in the large intestine.

This, though, is not what has got microbiologists excited at the moment. Just recently, a number of experiments have been done that show that the bacteria in your gut may have an ability to change your body in pretty radical ways. It looks like the gut microbiota in mammals can control not only weight but also mood and possibly even behaviour. Now, these experiments have all been conducted using mice, which means that there is a chance that the results do not directly apply to humans. Mice are a pretty good substitute for humans in this kind of science, though, and all of the work revolves around the use of special germ-free mice. These are bred in completely sterile environments, and over generations have come to the point where there are no bacteria anywhere on the mice or even inside them. The most immediate impact on the mice is that they have to eat a lot more food to maintain a healthy weight, which is a direct result of having no bacteria in their gut to digest fibre and produce nutritious SCFAs. Researchers from St Louis in the US then colonized germ-free mice with the gut microbiota of other ordinary mice. Another term for colonized is faecal transplant and, yes, that does indeed mean a poo transfer. What is interesting is what happens if you transfer gut microbiota from mice that don't have a normal weight. Put bacteria from an obese mouse into a germ-free mouse and it too becomes obese. Similarly, bacteria from an underweight mouse will make a germ-free mouse become underweight. There is clearly something that the gut bacteria are doing to the germ-free mice that changes their metabolism. Exactly what is going on is just beginning to be worked out.

A group based in Yale University in the US showed that it may be those nutritious SCFAs that are to blame. They fiddled with the gut bacteria of mice so that they produced more SCFAs and noted that this somehow turned on a whole slew of signalling systems in the brain that resulted in the release of the hunger hormone called ghrelin. Ghrelin is normally released into your bloodstream when your stomach is empty and increases your sensation of being hungry. So, when the mousey gut bacteria make too many SCFAs, you get hungry mice that then become fat. Perhaps even more fascinating are two sets of experiments done in 2016 by scientists in Cork, Ireland, and Houston in the US.

The Irish scientists showed that if you took gut bacteria from people suffering from severe depression and transplanted these into the germ-free mice, the mice became depressed too. Now, it may seem bizarre to call a mouse depressed, but there are ways to gauge the mental state of mice in a laboratory setting. Note that this was depressed human gut bacteria that changed the moods negatively in mice. On top of that, the

Houston group made young mice antisocial by giving them the gut microbiota of obese mice, and then made the juvenile mice social again by feeding them bacterial supplements. It looks like, in mice at least, not only is their weight partially regulated by their gut microbiota but also their moods and behaviour.

So, what does this all mean for humans? Well, while we can't yet be certain that the bacteria inside us are doing the same things, there is clearly a lot more going on than we thought. It has been known for a long time that stress in humans is linked to various digestive and bowel complaints. Irritable bowel syndrome, for example, often goes hand in hand with clinical depression. The assumption is usually that the depression is causing the irritable bowels, but it may be possible that it is the other way around and that the root of both problems lies with the bacteria in the large intestines.

However, there is one simple way you can change your gut microbiota. A change of diet can have a profound effect on the variety and numbers of bacteria in your large intestine. What is more, the effect can be really rapid. In a 2013 study from Harvard University in the US, it took just twenty-four hours from switching to a radically different diet for volunteers to show an equally radical change in their microbiota. Which brings us back to the things we eat and how it can have a profound effect on us in more ways than we previously knew. While it is probably not true that we are more bacteria than human cells, it looks like those bacteria play an essential role in our relationship to food.

The subtle science of killing bacteria

Given that the world is so permeated with bacteria, viruses and other bugs, on every work surface in your kitchen, on you, in your belly button and inside your gut, is it possible to protect food from these microscopic organisms? Any food grown outside of a sterile laboratory is, just like you, going to be covered in micro-organisms of one sort or another. Furthermore, we know that bacteria in particular will in time cause food to spoil, turning palatable nutrition into inedible rubbish. So, can you get the bacteria out of food? Yes, but there is a cost. If you want to completely remove bacteria and other micro-organisms from food there are several ways you can do this, but in so doing you will change either the taste or texture of the food. Consequently, most food-preservation techniques don't remove all the bacteria, they just reduce the numbers. It's usually a compromise situation where we balance the longevity of the food with alterations to taste and texture.

The first person to really show that it was possible to make something completely free of bacteria was the eighteenth-century Italian scientist Lazzaro Spallanzani. Ironically, he was not investigating food preservation but something more fundamental than that. By the start of the 1700s the development of the microscope had allowed us to see bacteria and other micro-organisms for the first time.

They were everywhere, it became apparent, and the question was where did they come from. The prevailing wisdom was that they spontaneously arose when non-living matter fused with a mystical spark of life. In 1768, Spallanzani set out to prove otherwise. It had already been shown that if you boil up meat broth you kill all the micro-organisms, but they always seemed to repopulate and grow anew. Spallanzani took broth and placed it in a sealed container, then dropped the whole lot into boiling water for an hour. The heat transferred through the container and into the broth, killing all the bacteria, and since the container was already hermetically sealed, there was no way for recolonization by micro-organisms. Thus proving that spontaneous generation of life was not possible. It was an experiment that others had tried before, but where they had failed was in creating a vessel that could be heated without the contents expanding and bursting out. However, Spallanzani was not interested, or it did not occur to him, that his proof that life did not spontaneously arise had an application in the kitchen.

It wasn't until 1810 that food preservation by heat treatment was cracked by a Frenchman, Nicolas Appert. Fifteen years earlier, in the closing years of the French Revolution, the military leaders of the recently minted republic (including the twenty-five-year-old General Napoleon Bonaparte) offered a substantial 12,000-franc prize for a new method of preserving food. Appert was a confectioner in Paris when the prize was first offered but set to work using technology he was familiar with from his home town, notably champagne bottles, and Spallanzani's technique. After initial successes, he moved to wide-necked bottles with corks sealed with a paste of lime and cheese. An odd mixture, but apparently one that survived the boiling treatment. His results were evidently impressive, his bottled peas being described in his 1810 book as having 'all the freshness and flavour of recently gathered vegetables', and he was duly awarded the prize money by the now Emperor Napoleon. But it was a different Frenchman, Philippe du Girard, who made the leap to working with metal containers. Since he couldn't get a patent in France, as Appert had that market sewn up, he essentially stole the idea and filed a patent in the UK through a London-based agent called Peter Durand, whose name is on the patent. Which is why Durand gets all the credit for inventing the tin can even though he did none of the work. By 1813, the first canning plant opened in Bermondsey, south London, but the canned food was expensive and only really used by the military and as a frivolous expense by those that could afford it. A particular impediment to the adoption of canned food was that the can opener was not invented until 1845. The instructions placed on early cans was that they should be opened with the aid of a hammer and chisel, a distinctly hazardous operation.

While those early cans were clearly revolutionary, as you could store food for years in them, the contents were hardly fresh. Despite the accolades in Appert's book, canned food is, by the very nature of the process, thoroughly cooked. That is not to say that the food is less than nutritious and tasty; in fact, for some items like grapefruit the canned version is more nutritious than the uncooked fresh alternative. The process of canning not only stops ripening in grapefruit, which uses up some of the fruit's vitamin content, but it also breaks down some of the fibre, making more nutrition available when you eat it. However, given that in Appert's case the food was cooked for an hour, I think it is safe to assume that his canned veg was probably overcooked.

The key to this is clearly reaching a compromise situation. Do you need to kill every single one of the bacteria? Can you get away with a less harsh process of heating? The answer is yes, you can, so long as you are willing to sacrifice some of the extended shelf life, and that's where pasteurization comes in. It's a process named after Louis Pasteur, who performed the definitive experiments in 1864. However, as is often the case in the history of technology, he was not the first to discover it. That honour should probably go to Japanese sake-brewing monks some 400 years before Pasteur. Either way, Louis ended up with his name attached to the process that is used to treat large amounts of the food we buy from our supermarkets. Take milk, for example, or any milk-based product for that matter. Raw milk, straight from the dairy, is pumped through a system of metal pipes that run in a hot water bath. By adjusting the heat of the bath and the flow rate of the milk you can ensure that the milk reaches a precise temperature for a precise length of time. For the UK, and

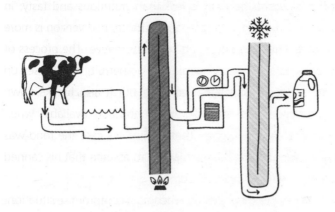

Pasteurization from cow to carton.

standards do vary a little around the world, the milk needs to hit 72 °C (162 °F) for fifteen seconds.

It's a specific requirement; the heat and temperature must be spot on. If you heat for too long or at too high a temperature you start to alter the taste of the milk. Whereas if the temperature is too low or held for too short a time, not enough bacteria are killed. So, how many bacteria are enough bacteria? And what is an acceptable number of bacteria in a carton of milk?

The bacterial content of milk varies enormously from farm to farm and from cow to cow. The bacteria can come from the cows themselves, either from dirty udders or in extreme cases from a teat infection called mastitis. Then there is the issue of the cleanliness of the milking equipment, the pipes, the storage vessels and the tankers used to transport the raw milk. All of which means that milk companies expect, on a good day, from healthy cows and well-run farms, about 10,000 bacteria per millilitre of raw milk (about 300,000 per fluid ounce). Which may seem like a lot, but remember that in an average human turd there are about 10 trillion bacteria weighing ten grams in total. So, it's all relative. Pasteurization of milk at just 72°C (162°F) for fifteen seconds will kill 99.99999 per cent of these bacteria. Which sounds comprehensively lethal, but let's break down the maths. There could be 100 million bacteria in a litre of raw milk (about 2.1 pints) and after pasteurization you would still have 10 bacteria left, give or take. So, what does that mean? Is that safe to drink? How long will a litre of milk last if it has 100 bacteria in it? To answer these questions, you need to consider what would happen to milk if it was not pasteurized.

Firstly, there is a chance that some of the bacteria in raw milk are pathogenic, or capable of causing a disease. These can range from the unpleasant, like food-poisoning bacteria, through to downright dangerous beasties like diphtheria, typhoid and even tuberculosis bacteria. These are all diseases of the past thanks to pasteurization, so we have forgotten how nasty they are, but they are really horrible and will kill about one in ten people infected. Clearly, swallowing down millions of potentially disease-causing bacteria is a very risky thing. Your body's own defence system will probably cope. However, if you are already a little under the weather, your immune system may just have too much on its plate and you could succumb to the bacteria in the milk.

There is another problem with having millions of bacteria in your milk: what all those bacteria eat. At their most benign, they will start to digest the lactose sugar in the milk, converting it to lactic acid. Initially this makes the milk taste sour but given enough time this acid will curdle the milk protein, turning it into a smelly, wobbly block of yoghurty stuff. On top of that, there are many other odorous molecules produced and I've not even mentioned the possible fungal spores and how mould will start to grow.

Clearly, we don't want any of that to have happened to the milk we consume. But if you can reduce the bacterial levels low enough you can significantly increase the potential safety and longevity of the milk. If you only have a few hundred or thousand bacteria, even if you are unlucky, only a small number of these will be pathogenic and your body's natural defences will cope with them. Similarly, those few bacteria will start producing lactic acid but only tiny amounts. On top of this, if you keep the milk cold in a refrigerator, you massively slow down the bacteria that are there. Unpasteurized or raw milk will last for maybe a day at room temperature. Refrigerate the same milk and it will last a week. Pasteurize it, though, and it will easily last two weeks and probably three. So, when it comes to killing off bacteria in our food, there is a balancing act we need to achieve between shelf life and taste.

Knowing when to eat your food

Around the globe, food manufacturers put a variety of helpful dates on the products we buy in our supermarkets. The exact wording varies a bit from country to country and what those dates actually mean can be significantly different. In the US, for example, most food will carry one of three types of shelf-life date: sell-by, use-by and best-before. The UK and the European Union have a similar set of date stamps, although the placing of sell-by dates is no longer recommended to food producers. The problem is that all these different date stamps cause an understandable confusion: can you eat food that is past its shelf-life date? Well, it depends which shelf-life date is being used and which set of regulations you are using. In the UK, the use-by date is the one you have to look out for. This means literally what it says, that you should have used the food, presumably by eating it, by midnight on the date shown. Once that date has passed the food is no longer safe for you to eat and it is recommended that you throw it away. The best-before date gives you more leeway; what this is telling you is that the manufacturers recommend that you eat the food by this date to appreciate it at its optimum condition. If you want to eat it after the date, go ahead but it won't be as palatable. The sell-by date is the one that used to cause the most confusion as this was just put there to help retailers keep track of how long their stock had been sat on the shelf. Sell-by was specifically not there to give the consumer any advice on when to eat the product. Which is why you don't see it any more in the UK. In the US, the equivalent labels have roughly the same meaning, but crucially none of them are officially safety labels and carry no legal weight behind them. The exact wording is also not fixed so the Food Product Dating code is more what you would call a guideline than an actual rule.

But how do manufacturers work out the dates and choose which foods get use-by and which best-before? The answer to the second of these questions is relatively straightforward. You need to consider what will happen when the food goes past a hypothetical shelf-life date. If the food product can become dangerous to eat, maybe because bacteria will have proliferated or poisoned the food, then you slap a use-by date on it. On the other hand, if the food just becomes less appetizing, maybe because it loses its crunch or becomes stale, then you put the best-before date on it. Foods in the use-by category tend to be wet foods like milk, meat or cheese, where bacteria will readily grow. Dry foods, on the other hand, usually get a best-before date instead.

As for how you determine the date, particularly the use-by date, well, that is a bit more involved. The key to this is the idea of what constitutes a minimum infectious dose of bacteria. This is the smallest number of bacteria you need to cause an infection in an ordinary, healthy person. Note that this dose

THE SCIENCE OF FOOD

of bacteria won't cause an infection in everyone, but it can cause an infection if you are unlucky. By which I mean that your immune system is not working optimally, maybe because you have a cold, or are sleep deprived, or particularly stressed about work. The minimum infectious dose varies from bacteria to bacteria, so you need to know what sorts of bacteria are likely to be found in the food you are trying to date stamp. Take, for example, raw chicken. A common contaminant of raw chicken is the salmonella bacteria that can cause particularly nasty food poisoning. The minimum infectious dose of salmonella is usually in the order of a hundred thousand bacteria. So, meat with less than this is deemed safe and the question now becomes, if you have a lump of raw chicken breast kept in the refrigerator, how long will it take for those bacteria to multiply to this potentially dangerous level? Once you have worked that out, it gives you a use-by date. Except the food producers err on the side of caution and use a date a few days before this, just in case the food was not stored at the optimum temperature.

To work out this date you can either take a load of samples of the food, stack them in a refrigerator and then test one each day to check the number of bacteria, or deliberately contaminate the food with the bacteria of choice and see how long it takes to grow to dangerous levels. But that is not how most shelf-life dates are determined as this is a laborious and expensive process. Instead, most manufacturers use a computer model specifically designed for a type of food. They input the various conditions of manufacture, transport and storage into the model and, using previous scientific data, it predicts the safe use-by date.

There is a considerable industry behind all of these calculations and determinations of shelf-life dates – an industry that makes it possible for us to safely buy our food from supermarkets with the certainty that it is fit for consumption. But there are many people who ignore the dates, regularly eat food after a use-by date has expired and are none the worse for it. Are the food producers being too cautious with their dates, presumably fearing causing harm and maybe suffering litigation? Are we needlessly throwing away food because it hits its use-by date, but is still good to eat? The answer to that is definitely yes, people do throw away good food that has hit its shelf-life date. But that is because the system has a built-in level of caution. It all comes down to risk and how you rate it (*see also* the five-second rule on page 133). Sure, you could eat that ham that is only one day past its use-by date, but are you sure it's not now contaminated with dangerous levels of *E. coli* bacteria that cause severe food poisoning? It's worth mentioning that food poisoning is not a small-scale or trivial illness. In the US, there are in excess of 50 million cases of food poisoning a year, 150,000 of these lead to hospitalization and over 3,000 deaths are caused. If the ham you are considering eating is within its use-by date and has been stored correctly, you can be sure that the chances of it containing dangerous levels of bacteria are so small that you need not worry. The point of shelf-life dates is to use the science of microbiology to take the guesswork out of knowing what is safe to eat. Which means that if we are agreed that throwing food away is a bad thing, then the sensible option is surely to only buy food you know you are going to eat and then make sure you scoff it before the date comes up. Although I appreciate that this is easier said than done.

The good, the bad and the fungi

So far in this section of the book I have sometimes glibly been using the deeply unscientific term 'bugs' to refer to the micro-organisms that share our world. I have also been pretty negative about the bugs in general, maybe with the exception of our gut microbiota. So, at this point I really need to set the record straight. Not only are many bugs incredibly helpful in the kitchen and in the preparation of food, but also not all bugs are bacteria. While it may be true that the majority of bugs we encounter in our day-to-day life come from the various families of bacteria, there are also fungi, tiny plants and even animals too small to see, although examples of microscopic animals that are both beneficial and used in food are vanishingly rare. How do the bugs help us make food?

By far and away the most common way bugs help make food is through the process of fermentation. You are probably aware that fermentation is the way we make all our alcoholic beverages, but it is the same process that underlies how we produce a huge range of food products. It's a surprisingly common undertaking that also gives us yoghurt, soy sauce, miso, fish sauce, crème fraîche, sauerkraut, kimchi and kombucha, and even salami is somewhat fermented. In addition, some primary ingredients need a fermentation step before they become usable; both chocolate and vanilla are essentially bland, lacking their distinctive aromas before the fresh beans or pods are carefully fermented. Each fermented food product has its own fermentation process that requires a specific micro-organism and specific conditions. However, at its heart is a very simple principle: a lack of oxygen.

Within all living organisms the energy needed to survive comes from a chemical process called respiration. The form of respiration we are most familiar with, since it is how we extract energy from our own food, goes like this. Step one is to take a glucose sugar molecule and split it into two identical pyruvate molecules, which are just intermediary molecules as far as respiration is concerned. This process alone releases a bunch of energy that our cells can use to survive. Then step two takes the intermediate pyruvate and adds oxygen, which breaks it down further and releases a whole load more energy. The key to this process is that while it is preferable to use both steps for maximum energy yield, you don't need to. Step one of respiration produces energy and some organisms choose to stop there, ignoring the second, oxygen-dependent step.

In some cases, they don't have a choice. Some bacteria living in waterlogged environments, where there is no oxygen in the first place, have become so used to sticking with just step one that oxygen is positively poisonous for them.

There is, however, a wrinkle to this. To get step one to happen – breaking your glucose molecule into two pyruvate molecules – you need to use another chemical called nicotinamide adenine dinucleotide, or NAD for short. A molecule of NAD has two forms, NADH and, if you take off the hydrogen, NAD^+. It is relatively easy to convert between the two and NAD is endlessly cycled between NAD^+ and NADH in all the cells in our body. When you do step one of respiration you flip a NAD^+ to NADH. So how then do you turn the NADH back to NAD^+? You could just use up some of the energy you created in the first place, but if you are stopping at step one and not doing the step that needs oxygen, you can use fermentation.

This is where the scientific definition of fermentation comes in. If you take the result of step one of respiration, notably pyruvate molecules, you can use them to turn NADH back to NAD^+. There are two ways to do this and each produces a desirable waste product. The simple option is a direct conversion from pyruvate to lactic acid. Alternatively, you first snip off a carbon and two oxygens, making carbon dioxide gas, and the remaining bit of the pyruvate is then converted to alcohol, specifically ethanol. If an organism is going to use oxygen-free respiration, it is going to produce either lactic acid or ethanol as a waste product. Which process it uses depends on the organism. Bacteria tend to favour the lactic acid option, while microscopic fungi, like yeast, go for the ethanol route. Which begins to explain the two main types of fermented food.

If you ferment with an organism that produces lactic acid you end up with things such as yoghurt, sauerkraut and kimchi. The most common bug responsible for these foods is a bacterium called *Lactobacillus*. If you put *Lactobacillus* in an oxygen-free environment, it will switch from full-blown respiration to lactic acid fermentation. As it grows it begins to churn out lactic acid and that acid can be used in a couple of different ways. If you introduce lactic acid to milk it causes the proteins in the milk to change shape: they denature (*see* page 28) and turn into long spaghetti molecules. These then tangle up with each other and turn the liquid milk into a solid, set lump most commonly known as yoghurt. Alternatively, if the starting material is a vegetable, the end result is an acid-pickled food like sauerkraut (German pickled cabbage) or kimchi (a Korean dish, usually of pickled cabbage and radish).

On the other hand, if you are looking to produce alcohol or carbon dioxide gas, you need to add a fungus to your food. The most popular is the single-celled *Saccharomyces cerevisiae* or yeast. If you put yeast in a mixture with lots of food, either sugar or starch, it will take the easy route and only perform step one of respiration, fermenting the results to make carbon dioxide and ethanol. If you start with something like grape juice, the sugar is converted to pyruvate and that in turn becomes the alcohol that makes the juice into wine. What may be less obvious is that the exact same process is how we make bread. The yeast ferments the sugars in the starch (*see* page 70) and makes carbon dioxide gas, which makes the bread rise. In each case, the other product of the fermentation is also produced. Wine produces carbon dioxide, but it is allowed to escape. And when making bread some ethanol is made, which usually evaporates in the baking process. Something like champagne uses both waste products of the yeast. The waste alcohol makes the grape juice into wine and the carbon dioxide is trapped to give it the fizz.

Both types of fermentation are important in the production of huge amounts of our food. Most rely on just one type of fermentation but a few mix them together. Sourdough bread relies on a combination of yeast and bacteria to produce a distinctive result. The bacteria begin to digest sugars that the yeast can't cope with, producing lactic acid. This gives the bread its distinctive tang, but also provides a food source for the yeast, which makes carbon dioxide through fermentation and the bread rises.

Fermentation has been around since the Stone Age, for probably 10,000 years, and is the earliest and most extensively used form of food processing. Without it your supermarket shelves would be decidedly bare.

Eating bugs

Fermentation is not the only way we use micro-organisms to make our food. There are a few rare cases where it is the bugs or micro-organisms themselves that we eat. The two common examples here in the UK are both based on microscopic single-celled fungi. The first is pretty straightforward but also uniquely British. Marmite is a thick, black and salty paste that some British folk like to spread on toast. It tends to split the population into those that love it and those that can't abide it. It would also seem that this is one of those things with such a peculiar flavour that you need to be brought up on it from childhood to have any chance of liking it.

It was invented by a nineteenth-century German professor of chemistry called Justus von Liebig. He found that if you took yeast that had been used to ferment beer, it could be turned into a concentrated black paste that smelled and tasted distinctly meaty, even though it was entirely vegetarian. It was mostly ignored until 1902 when the Marmite Food Extract Company set up a factory in Burton upon Trent in the UK to make the stuff. They chose Burton as their base so that they could use the waste yeast from the Bass brewery just next door. The product was named Marmite after the French-style ceramic pots it was originally sold in. Since Liebig's time, science had discovered that certain chemicals, dubbed vitamins, were vital for life, and Marmite turned out to be packed with these chemicals. Consequently, it quickly became a popular, cheap and nutritious food within the UK.

The production process for Marmite is ridiculously simple because the yeast cells do a big chunk of the work. The waste brewer's yeast is mixed with lots of salt and allowed to stand for about twelve hours. The salt has a peculiar effect on the yeast, causing the tiny fungal cells to self-destruct. The salt kicks off a cascade of chemical and biochemical reactions that kill the yeast cells and release enzymes that digest resulting debris. It's all a bit gruesome, but the result is a creamy brown, thick soup. This is then concentrated by allowing it to flow down a long series of heated pipes. It goes in the consistency of cream and comes out like treacle, thick and black. But unlike treacle, it's very salty.

They are a few different versions of yeast extract produced around the world. The Australians have Vegemite, which has added celery and onion concentrate but is still basically yeast extract. The New Zealanders have a product also labelled as Marmite, although to the connoisseur it has a sweeter and less salty taste. Differences aside, all are made from yeast, a single-celled organism.

The newest kid on the block in the field of microscopic fungal-based foods is Quorn, a vegetarian, protein-rich, meat-substitute product that can now be bought in supermarkets around the world. Quorn is named after the small British village just north of Leicester where it was first manufactured as a commercial product in 1985. But it was back in the 1960s that work began to identify new sources of protein. It had been predicted that by the 1980s the world would be suffering from a global food crisis. Thankfully this never materialized, but it did spur food scientists to start looking around for alternatives. In 1967 a fungus was found in a sample collected from a local field by scientists working at the Rank Hovis McDougall Research Centre near Marlow, to the west of London. That fungus, called *Fusarium venenatum*, proved to be ideal. Not only could it be persuaded to grow in huge tanks but, rather than making little round blobs like yeast, it produced short, thin filaments. Those filaments turned out to be about the same dimensions as individual muscle fibres. Consequently, if you press the filaments to form a block, it has an uncannily meaty texture.

It took eighteen years to perfect a way to grow the fungus in bulk and produce a palatable product from the microscopic fibres. These days Quorn and similar products have been made into a whole range of meatless products from lumps that resemble chicken-breast meat to slices of bacon. It is probably worth noting that while the global meat alternative market in 2016 was worth in excess of $4 billion, it is not vegetarians that are eating it. After all, if it looks likes meat, tastes like meat and has the same texture as meat, it's probably the last thing a vegetarian is going to want to eat. Some 90 per cent of Quorn, for example, is consumed by omnivores who are just trying to cut down on the amount of actual meat they eat. However, whoever the consumers of Quorn are, it remains a remarkable product set apart from the vast majority of our food as, like Marmite, it's made from a fungus you can't even see.

The Future of Food

Food pills are not the future

I've always been a bit of a science nerd and along the way I predictably picked up a love of science-fiction literature. It occurred to me that while some of the tropes of culinary science fiction have come to pass, like artificial meat (Quorn on page 160 and synthetic burgers on page 167), there is one idea that has resisted being realized. Food pills that can substitute for a full meal have been a staple diet for science-fiction writers for a very long time. I remember seeing them as a child in cartoons, on reruns of *The Jetsons* from Hanna-Barbera, and in comics Dan Dare was partial to them. But you need to look much further back to find the first mention of food pills.

In July of 1879, a journalist by the name of Edward Page Mitchell had a short story published in the *Sun* newspaper, a now defunct New York publication. The story was called 'The Senator's Daughter' and was a futuristic take on a Romeo and Juliet theme. But there is a subplot of social comment running through the story that features compact food pastilles, or food pills, tackling the issue of how to supply sufficient food to an ever-growing global population, something that was of concern in the 1870s. In the story, an invading army decries meat on the grounds that it is wasteful and unethical, but some go a step further. These radicals apply the same logic to vegetable matter. Hence the need for a completely artificial food source made directly from primary ingredients, notably elemental carbon, nitrogen, hydrogen and oxygen.

From there the food pill becomes one of the standard science-fiction tropes. It appears again in 1893 in an essay by Mary Lease, an American suffragette writing for the World Fair held that year in Chicago. Lease took a different view on the food pill, claiming that with the aid of science, by 1993 we would all be eating synthetic food and the food pill would emancipate women from the drudgery of cooking.

Thereafter the food pill is a regular in the genre until about the 1980s when it fell out of favour. But why? It turns out that while we do have a pretty precise grasp of the nutritional requirements of the human body, there are some limits placed on food by physics. The daily recommended energy intake for an adult woman in the UK is 8,400 kJ (that's kilojoules as the unit of energy, or about 2,000 kilocalories). Of all the things we can eat, pure fat has the most energy crammed into the smallest space. It is at least twice as energy dense as sugars, proteins or complex carbohydrates. A cube of solid fat, say coconut oil for arguments sake, that is 6 centimetres (about 2½ inches) on each side would provide you with enough calories for a day. Imagine trying to eat a block of fat a bit bigger than a Rubik's Cube. That is the smallest volume you can fit the needed energy into. So, you would roughly have to ingest about 200 gargantuan pills to get your recommended energy intake; that's sixty-six pills every breakfast, lunch and dinner. If you are male, you're looking at over eighty pills every time you sit down to eat. The physics or maybe the chemistry of energy density results in each meal being a plate full of pills, which is a long cry from a single compact food pastille.

That is, of course, only considering the total energy you need. On top of food that can be converted to energy we need proteins to turn into the proteins our own bodies are made of, fats for cell membranes and then there are all the micronutrients like iron, potassium, sodium and the vitamins. Add this together and you are going to end up with a mound of pills on the plate. Even if you could gather all these nutrients into pills, there is something else that is missing and fundamental to not only our health, but the health of the bacteria living inside us (see page 137). Fibre in our diet helps ease the movement of things through our digestive tract. If you ate only what you needed, with no waste at all, then you would have no bowel movements. Which would mean that your large intestine would be entirely empty. You would lose all the hormonal control coming from the gut microbiota (see page 137) and, worse, the lining of the gut would begin to break down, disrupting your immune system and also leading to a potentially fatal bacterial infection of your blood or internal organs. The bottom line, pardon the pun, is that to be healthy we need to poo. It's this that really makes the food pill a science fantasy. Our digestive tracts have evolved to have food and waste products in them. Reduce our food intake to a pill and, as far as we know, the effect on our physiology would be quite possibly fatal.

The meat that isn't

If we want a more satisfying meal than pills that still takes on some of the aspects of science fiction, the carnivores among us do at least have exciting options in the next five years or so. The latest advances in synthetic meat may be bringing you turkey grown within your own home and not a feather in sight.

The idea of growing meat in a laboratory setting is far from new. In the 1932 edition of *Popular Mechanics* magazine, a gentleman by the name of Winston Churchill, before he was made the prime minister of the UK, wrote that 'synthetic food will, of course, also be used in the future'. He also said that we would be growing pieces of chicken to eat, rather than the 'absurdity of growing a whole chicken'. It must be said that Churchill, as so many science-fiction prognosticators, was a bit overenthusiastic with the rate of progress. In his article in the magazine, he was predicting the state of the world in 1982, fifty years from its publication. It was not until 2003 that the first truly synthetic meat was grown and eaten, but this was not done in the name of science alone.

Artists Ionat Zurr and Oron Catts grew a tiny steak of frog muscle cells, which was exhibited and eaten at an art exhibition held in the city of Nantes in France. Cells were harvested from a live frog and grown in an incubator in the art gallery for the duration of the *L'Art Biotech* exhibition. The donor frog, and a few of its mates to keep it company, lived in an aquarium alongside the growing synthetic meat. At the close of the exhibition, the artificial frog steak was cooked nouvelle cuisine style, presumably because it was tiny, having been first marinated in Calvados brandy and then fried with garlic and honey. This pioneering, artificial meat eaten by a human was not altogether delicious. It was described as being like 'jellied fabric'. If you are wondering what happened to the live donor frog and chums, the story ends well for them as they were released into a pond at the local botanical gardens.

The frog-steak art installation did not raise much interest in the public eye, mostly because the product was patently nothing like a real meat steak, from a frog or otherwise. But ten years after the Nantes froggy food, a believably meaty artificial burger was consumed. It was in August 2013 that Professor Mark Post, of Maastricht University in the Netherlands, unveiled his synthetic beef. It was cooked and eaten in front of a live news conference in London and was described as being close to meat but not as juicy, which is pretty impressive for a first product. Sadly, the cost of this particular beef burger was a distinctly juicy €250,000 (which at the time converted to £216,000 or $331,000).

The process of creating the burger started by extracting cells from cow muscle. Conveniently for science, muscle has a natural ability to grow and repair itself. Living within all muscles are cells called myoblasts that will, when the muscle is strained, respond by multiplying and fusing to form new muscle fibres. This is why going to the gym to lift heavy weights does something. Your muscles are exercised, the myoblasts get to work and you get bigger muscles. Professor Post and his Maastricht team extracted myoblasts from a cow muscle biopsy and grew them in special nutrient broth until they had billions of them. They then did something that enabled them to get around the 'jellied fabric' issue. They put the myoblasts into tiny wells a few millimetres across containing an even smaller central pillar. The myoblasts took it upon themselves to arrange into muscle fibres around the pillar and began to spontaneously contract, presumably giving the pillar a tiny hug each time. This ability to exercise stimulated the myoblasts to further develop into tiny rings of muscle tissue, just about visible to the naked eye. It's these rings, some 40 billion of them, that were cut open to create tiny strips and packed into a Petri dish to form the burger.

The main problem the burger had, and this was called out by the taste testers, was a lack of fat. Since the synthetic meat was derived from just myoblasts, and they only grow up to make muscle cells, there was absolutely no integral fat. If

How to make meat.

you watch the promotional video of the cooking and eating of the burger, the chef bastes the burger with lashing of butter as it fries to moisten the otherwise potentially rather dry beef (*see* page 105 for why fat in meat makes a difference). However, there is no doubt that it is perfectly possible to create a convincing artificial meat.

Since then there have been further developments and Professor Post now has competition. The most exciting of which is an American company based in California somewhat confusingly called Memphis Meats. Their initial breakthrough was the creation of the world's first lab-grown meatball, which was unveiled with much fanfare in the media. Now, I don't want to appear to be too critical, but isn't a meatball just a burger in a different shape? That aside, the folks at Memphis Meats have recently unveiled the world's first synthetic chicken and duck products, which mimic the consistency of breast meat from the respective species. At the big press announcement for their new products they cooked up a crispy-coated fried chicken and the duck was pan-fried in an orange sauce. All the tasters said it was delicious and it certainly looked good.

It's clear that the technology and science to create convincing meat from cultured animal cells is now well established. What has not happened so far is scaling this up to the point where it can be used to make a cost-effective product. However, this challenge is essentially the same as has been achieved with Quorn since the 1980s (*see* page 160). To mass-produce cultured artificial meat you need huge 2,500-litre (660-gallon) fermentation tanks to produce the starting cells. What is a bit trickier is the step that produces the muscle fibres, but Professor Post and his Dutch team are nearly there. The latest figures released by them claimed that they had the price of a burger-sized piece of artificial beef down to around €8 (which at today's exchange rates is £6.90 or $8.60). Still too much to be commercially viable, but not far off.

The benefits of artificial beef or synthetic chicken are significant. A big bonus is that it removes any welfare issues you may have with the production of meat products. On top of this, it frees up large amounts of agricultural land. You no longer need to pasture animals, nor do you need to dedicate land to growing crops to feed them, which is a particular issue in the case of cows. It is also a more energy-efficient process overall when you consider all the many energy requirements for both products. However, it is not entirely a bed of roses. Two issues remain with artificial meat. First there is an uncertainty in the public's perception of these foods. The term Frankenfood first appeared in 1985 but didn't really start taking off in the media until the turn of the century. It's a phrase that makes me wince every time I see it as it is clearly implying that the food is monstrous like the literary monster created by Mary Shelley's Dr Frankenstein. My guess is that the benefits of this technology and synthetic meat products will help change opinions and overcome squeamishness. Especially if they keep publicizing them by cooking up delicious-looking dishes at their press conferences.

There is another problem more fundamental to the process: where do you get the nutrients needed to grow your meat muscle cells? The sugars, vitamins and micronutrients needed are not an issue: there are plenty of plants and yeasts that can help with that. The problem is the protein. You need to feed the growing myoblasts with protein and, at least initially, researchers in this area have used animal sources of protein. The whole point of making artificial meat is to unhitch the production of a primary protein source from the need to grow animals on farms.

Making enough protein to feed the world

The problem of where our protein comes from is a much bigger issue than just what to do with synthetic meat. Protein is one of the essential nutrients our bodies need. It is used primarily for the growth and maintenance of our bodies, but in an emergency it can be broken down and used as an energy source as well. When you start looking at our nutritional needs for protein it gets complicated. The current nutritional guidelines say that you should consume three-quarters of a gram of protein per day for every kilogram you weigh. So, if you are 60 kilograms (average female), that works out at 45 grams of protein, and for 70 kilograms (average male) it is 55 grams (1.6 oz for 132 lbs weight and 2 oz for 155 lbs weight), but there is a wrinkle to this. What confuses the issue most is the way that it has been taught and explained in the last few decades. Terms like good and bad protein or first-class and second-class protein falsely imply that there is a hierarchy in protein. The problem is that protein is not one thing but twenty.

Protein is made up of long chains of molecules called amino acids and there are twenty different types of amino acids in humans. It is not the protein that we need in our diet so much as the amino acids. When you consume protein, your digestive system breaks it down into its component amino acids, which are then absorbed and used to build the proteins we need in our bodies. But we only need some of the twenty types of amino acids. There are nine essential amino acids. If you don't get enough of these you will sicken with a variety of malnutrition diseases and in extreme cases it can lead to death. The other eleven amino acids are, so long as you are generally healthy, non-essential or dispensable, the reason being that our bodies can manufacture the dispensable amino acids from the essential ones.

One of the persistent myths of nutrition is that meat contains all the essential amino acids and is thus a complete source of protein, whereas plants do not and are incomplete. This is now understood to be just plain wrong. Plant protein sources are also complete and contain the full range of essential amino acids. This doesn't just apply to things such as beans, nuts and seeds that are traditionally seen as good protein sources. It's also the case for things like cauliflower, spinach or lettuce. Admittedly, there is not much protein in these veggies, but what is there is as complete as beef-steak proteins. There are also a few plant-based protein sources that have lower than ideal levels of some amino acids. Maize or sweetcorn, for example, contains slightly less of an amino acid called lysine than we actually need. Or to be precise, the proportion of lysine in the protein in sweetcorn is a bit low. Sweetcorn protein is generally only 2.5 per cent lysine, whereas humans need about 5 per cent of our protein to be lysine. Now, you can compensate for this in two ways. Either you eat a pile of extra sweetcorn, increasing your overall protein intake and thus bumping the total amount of lysine you consume. Or you could eat some beans with your sweetcorn as they are rich in lysine and will complement the deficiencies of the sweetcorn. Which is exactly what traditional Central and South American cuisine does, where beans and corn are a staple with one rarely being eaten without the other.

So, now that we have cleared up some of the subtleties of protein nutrition, how does this help tackle the problem of where humanity is going to get its protein from? It may not occur to you that this is even a problem – after all, the diet of developed countries is usually laden with more protein than is healthy – but you need a global view on this. The population of the world is currently around 7.5 billion and it is estimated that 2 billion of these people, just over a quarter, suffer from some kind of malnutrition. By the year 2040, the global population is predicted to break the 9 billion mark. The worry is that we are reaching the limit of our ability to produce food and that, as the population grows, an even greater proportion will find themselves suffering from malnutrition. Clearly, this is a complex problem but at least part of it is where we get our protein from, as bulk calories from starch are probably going to be less of an issue.

It is possible to measure the efficiency of different protein production systems by looking at things like land or resource usage. So, how much farmland does it take to grow the cow that is used to make the beef burger, and how does that compare to pork, lamb, chicken or what about soybeans?

Clearly you can't make a beef burger out of soybeans, but you can compare the quantity of edible protein produced. The results don't look great for the meaty options. Cows take about seventy times as much land as soybeans to produce a kilogram of protein. The difference is less for pigs and chickens, but it is still in the region of twenty times the land needed for a high-protein vegetable source. On top of that, to grow vegetables you need a fraction of the water and have far less greenhouse-gas emissions. On that last point, cows and sheep are particularly problematic as their digestive systems produce vast quantities of gas that they are constantly burping up. Admittedly some farting does also occur, but 90 per cent of the gas created comes out of their mouths. In the UK, cows alone contribute 3 per cent of our total greenhouse-gas emissions. What's worse is that the gas emitted is methane, and this is at least twenty times better at trapping the heat from the sun than the most common greenhouse gas, carbon dioxide.

So, does this mean that if we want to feed the world with enough protein we are all going to have to stop eating meat? Growing plant-based protein sources is much more efficient, but we need to then make sure the balance of amino acids is correct. With cultured meat or the use of microscopic fungi (see page 154), the calculations being done right now put it on a par with vegetable protein sources when it comes to land use. The energy required is still higher, but it is a technology in its infancy and from an energy perspective it is much better than growing whole animals. The latest challenge for the synthetic meat pioneers is to develop plant-based protein nutrient soups in which to grow their muscle cells. While it is a challenging task to find an efficient way to do this, there is nothing theoretically difficult here. It's just going to take some hard work to create the fully industrialized and scaled-up system.

The other possible way forward is that we turn to non-traditional food sources, at least as far as Western diets are concerned. The most obvious of these is that we start eating insects like mealworms, caterpillars, beetles and crickets. Immediately this proposal strikes a brick wall. When I suggest that you try eating a caterpillar, your reaction is probably one of disgust. This is entirely to be expected as, within Western culture, there is what is officially known as the 'yuck factor'. In 2013, the Food and Agriculture Organization of the United Nations produced a very thorough, very long and very boring report on entomophagy, or eating insects to you and me. In this, they identified a few major problems when it came to getting people to scoff insects and the yuck factor was the biggest of these. Culturally, we are programmed to not want to eat insects. We see them as vile and unfit as a source of food, but there is no logic to this. After all, prawns, crabs and lobster are considered a delicacy served in top restaurants around the world and they are all crustacea, closely related to insects. On top of that, about a quarter of the world's population already regularly eats insects. It's just that the people that do are not part of the Western culinary tradition. Ironically, many of the insect-eating cultures around the world are abandoning the practice as they seek to imitate the more affluent and desirable habits of developed nations. The great thing about insects is, like synthetic meat, they don't need many resources to grow. They can be farmed in much smaller spaces per kilogram of protein produced. They take less water, energy and emit minimal greenhouse gasses. There are a few concerns to do with allergies and the impact on wild insect species, but overall the picture for using insects as a protein source looks good.

How then do you convince a person to bite down on a juicy caterpillar? The simple answer is that you don't. At least you don't give them a caterpillar to eat in the first instance. It is going to be a very gradual process of introducing us to the idea that insects are yummy. One of the first steps will be adding insects to bulk up the protein content of animal foods. It would be a fairly easy thing to do and would at least get us used to the idea of insects in the food chain direct to our own mouths. From there, the obvious step is to introduce ground-up insects as an ingredient. This can then be added as a high-protein bulking agent to processed foods. But if you want to tap into the full geek culture, I've seen 3D-printed food snacks made from what was called mealworm flour. Essentially, it's just ground-up dried mealworms made into a paste, flavoured and then squirted through a nozzle to make whatever shape you want, so long as it's flat. The resulting splat of white goo is allowed to air dry and you end up with a crispy snack food containing a highly nutritious 50 per cent protein. It's clearly a gimmick, but by adding that 3D-printed novelty people are more likely to give it a go. Once ground-up insect powder is established on our list of acceptable ingredients then it is just a matter of time before toasted crispy crickets with garlic and lime or fried silkworm pupae on a stick make an appearance in our food shops.

If you don't fancy entomophagy then you could have a go at eating microalgae. As the name suggests, these are very small varieties of algae, which is the same stuff as seaweed. Confusingly, microalgae also refers to cyanobacteria, which aren't algae at all but a whole different kettle of fish. Confusion aside, what we are talking about are microscopic, single-celled green organisms that thrive in water and gain their energy the same way plants do, notably from photosynthesis. They are very similar to the organism that is used to produced Quorn (see page 160), but unlike the Quorn microfungi, microalgae only need sunlight, water and a few essential minerals to grow. At the moment, the most common of the microalgae used as a food is one called spirulina, which grows as elegant but microscopic spirals dotted with green blobs. Despite this being a new and exciting food product for our Western diets, humans were eating it long ago. Apparently, the Aztecs were scooping it out of Lake Texcoco and making dried cakes of it. It was also traditionally harvested from Lake Chad in central Africa and used to make *dihé* broths, and it's from here that it entered the Western diet. These days it is sold as a dark green powder and contains an impressive 60 per cent protein, which is more than other big vegetable hitters like soybeans

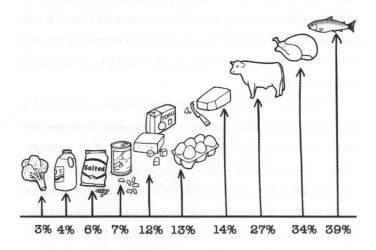

3% 4% 6% 7% 12% 13% 14% 27% 34% 39%

(56 per cent protein dry weight). Unfortunately, even its greatest advocates will admit that spirulina powder tastes vile. Imagine the taste of a really strong kale, but concentrated with an extra dash of bitterness and a hint of rotten seaweed. As I say, an unfortunate flavour, which is why it is currently only really used as a food supplement and is added into smoothies and the like. On top of that it makes your poo turn green. Despite its drawbacks, it remains an incredibly good source of protein that can be produced in large quantities on a very small amount of land and has potential in the future.

If we want to feed the growing population of the world with enough protein, we are going to have to uncouple our diet from its dependence on meat as its primary source of our essential amino acids. On top of that we are eating too much protein and the amount is growing every year. In 2014, the average meat consumption for adults in the US was 90 kilograms (198 lbs) per year. That translates to 247 grams (8.7 oz) of meat per person every day. Now, meat is only about one-quarter pure protein, so the average American gets 62 grams (2.2 oz) of protein from their daily meat eating. Add to this protein from dairy products, grains, vegetables and supplements, and you are looking at over twice the recommended allowance. So, the solution to the protein conundrum is twofold. Eat less of it and try a few of the alternatives. Although, as someone who is not a great fan of cabbage flavours, I'm personally not recommending the spirulina powder.

Nuts to allergies

If we are going to all these lengths to create new sources of protein to feed the world, it makes sense that we also need to ensure that the protein we already have access to can be eaten by everybody. For some people, it is not just how much and where your protein is coming from that is an issue. For them the world of food is even more complex and not a little fraught with danger. If you have a severe allergy to something like peanuts, then you need to be extremely careful what you eat.

Even a tiny amount of peanut can set off a very severe allergic reaction. Can the future of food overcome these allergies?

For many of us, allergies are an everyday occurrence, especially in the spring and summer when there is pollen in the air. I'm a hay-fever sufferer, and for me my symptoms kick in at the end of spring in May, go through the summer and usually clear up in August. At worst, it is unpleasant and can leave me feeling a bit miserable, but it is also completely manageable so long as I remember to take a single, minuscule hay-fever tablet each morning. The point is that my hay fever is an inconvenience at worst, and that is because my allergy does not induce anaphylaxis, a particular type of very severe allergic reaction.

The reason why some of us become allergic to things and others don't is still not really understood. Part of the problem is that the biology behind an allergic reaction is all to do with our ridiculously complicated immune system. Fundamentally, when you have an allergic reaction to something, it is your own immune system getting the wrong end of the stick. To protect our bodies from all the possible pathogenic bacteria, viruses and invading fungi, we have evolved a very elegant, very efficient and very complex immune system. It can do remarkable things: detecting, isolating and destroying anything that is not supposed to be part of us. One of the most remarkable things about it is that is has a memory. Once your immune system has been exposed to and has dealt with a virus, for example, it remembers the virus and any subsequent exposure is dealt with before the virus can gain a foothold and make you ill. This is the basis of how vaccinations work, allowing us to eliminate terrible diseases from the world. But as with any complex system, there are inevitably weaknesses and loopholes.

In the case of allergies, as a young child your immune system is exposed to a common and harmless substance. Then rather than ignoring this inconsequential material, your immune system gets all fired up and decides that it has just discovered a dreaded pathogenic invader. It responds by throwing all of its resources at it, which results in some collateral damage to you, like itchy skin, puffy eyes and so on. Your immune system then makes careful notes and remembers what this substance looked like. Consequently, when you are then exposed to the substance once more, a repeat allergic reaction takes place. So, for me I must have inhaled grass pollen as a baby. My immune system decided not to ignore it like most sensible immune systems would, but instead set me up for a lifetime of sneezing and complaining about walking near hay fields. What makes some people develop an allergy is really not understood, but there is definitely a genetic element to this. Which makes sense as we know that we all have a unique immune system that is inherited partly from each of our parents. It also looks like only some things are capable of setting up an allergy; we don't understand why this is, but peanuts are particularly good at it.

What makes peanut allergy worse is that the reaction it produces can be far more extreme than my mild hay fever. This extreme version of an allergy is called anaphylaxis and can be deadly. If a person with a severe peanut allergy eats even a morsel of peanut, within minutes or even seconds they will feel their throat beginning to swell. Combined with this, their airways constrict, making breathing difficult. Worse is to come as they suffer a massive drop in blood pressure as all their blood vessels suddenly expand. There are also some unpleasant effects like vomiting and diarrhoea, but the real issue is that since you are struggling to breath, you can't get enough oxygen to your brain and muscles. On top of this, the low blood pressure gives you heart failure, and that's why it is possible for someone to die from anaphylaxis. The treatment for such severe cases has to be administered very rapidly and is usually in the form of an injection of adrenaline, or to give it its proper name, epinephrine. This very quickly constricts your blood vessels and pushes your blood pressure back out of the danger zone. People who know they could potentially suffer such an extreme reaction usually carry around an automatic adrenaline injection system known by the brand name EpiPen. They are ridiculously simple to use. You hold the EpiPen in your fist, pop the cap and thump the patient on the outer thigh with the exposed end of the pen, which automatically injects the life-saving drug. A trip to the hospital emergency room is then always recommended.

Now, here is the real problem: it looks like the number of people who suffer from food allergies, including a reaction to peanuts, is on the rise. At least, we think that is the case. It turns out to be very difficult to measure this as over the last ten years there has been a massive increase in public awareness of food allergies. It is also mostly self-diagnosed and people often mistake food intolerance for allergies, which are completely different things mediated by different processes. However, it has been possible to gather some hard data. Between 1990 and 2004, the number of hospital admissions in the UK for anaphylaxis, caused by anything from bee stings to peanuts, has gone from five per million of population to thirty-six per million. Admissions for less severe but specifically food-related allergies went up fivefold in the same period, and if you look at just admissions of children, it was a sevenfold rise. Clearly, something is going on.

It may be to do with the way we are cooking our food. Peanuts are a staple food in lots of countries like the US, Australia, the UK and China, but peanut allergy is not nearly as prevalent in China. The biggest difference would appear to be the way peanuts are cooked. In China, peanuts are often boiled and mashed or fried, whereas in all the countries that have lots of peanut allergy they are almost always roasted. It is quite possible that the act of roasting is doing something to the allergy-inducing chemicals in the peanut. It may also have something to do with how and when a child is exposed to peanuts. One study from a group in Bristol in the UK found a correlation between the development of a peanut allergy and skin creams that contained peanut oil and trace amounts of peanut protein. But that is only one study and the current leading explanation is known as the hygiene hypothesis.

This states that allergies have increased due to a drop in childhood infections. At the moment, there is a substantial amount of population data to back this up, by which I mean that populations of people with lots of food allergies tend to have fewer childhood diseases and gut parasites. However, correlation does not imply causation. While this theory does have a potential causative mechanism buried in the fantastically complex workings of our immune system, what it lacks is a definitive experimental proof of any kind.

Where then does that leave us tackling the growing epidemic of food allergies, both mild and severe? How can we feed the world when some of the most prolific plant-protein sources, such as soy and peanuts, are becoming deadly allergens to an increasing proportion of the population? Two solutions are currently being investigated. First, you have immunotherapy, which is basically aversion therapy for your immune system. The idea is that you gradually introduce the substance you are allergic to into your body and slowly increase the quantity over time. The hope is that your immune system becomes reprogrammed to ignore the thing you are allergic to. It works for some people but can be a dangerous treatment, especially since we are not really sure how or why it may work. In one tragic accident in a 1996 immunotherapy trial, a patient allergic to peanuts was mistakenly injected with a large dose and literally died in seconds with the needle still in them.

A more promising approach may be to take the allergy out of the peanut. A recently formed start-up company, Aranex

Biotech, based at Warwick University in the UK, is taking peanut plants and using the latest genetic engineering tools to turn off the genes responsible for the allergy-producing proteins in the peanut. So far so good, but all these proteins are used by the plant as part of the food storage system for the seed. We don't yet know what a plant without these essential proteins looks like. It's hoped that the other, non-allergy-inducing storage proteins will be increased in compensation. If it works, you would have a completely allergy-free peanut that is safe for anyone to eat. Not only that, exposure of children to this sort of peanut would avoid them developing an allergy in the first place. Based on initial interest of a few companies, it looks like the first place you may see these peanuts will be in chocolate bars. After all, chocolate and peanut is a match made in heaven that should be available to everyone, allergy or not.

Super-charging plants

Ultimately, all of our food comes from one chemical reaction: photosynthesis. The vegetables we eat are clearly made that way but so are any animal products we eat. The milk you drink, the eggs or pork you eat are all from animals fed on plant-based feed. If you take that thought to its ultimate conclusion, you and I and virtually every bit of organic matter on the planet can trace its source back to photosynthesis. So, to say it is a pretty important reaction is more than just mild understatement. What if we could make photosynthesis, the source of all things organic and specifically our food, even better than it already is? If photosynthesis was more efficient, crop plants could produce more crops and a field of corn would feed more people. Now, you may be thinking that we have no chance of that – after all, plants have been evolving for some 600 million years. Photosynthesis itself first appears in the history of life even further back, long before land plants appeared, some 3.5 billion years ago. But here's the rub: evolution is not about maximizing efficiency, it's about survival of the fittest and what constitutes fit may not be the most efficient.

The full chemical pathways of photosynthesis are complicated, convoluted and daunting. But you only need to know about the bit at the heart of the reaction to see the potential for efficiency improvements. The basic process of photosynthesis starts with the plant capturing the energy of sunlight using the green pigment chlorophyll. That energy is then used to make a chemical compound inside the plant that is basically a chain of five carbon atoms with some bits hanging off it. The five-carbon molecule has the tongue-twisting name of ribulose bisphosphate, or RuBP to its friends. The next step, and don't look away now as this is the crucial bit, is the addition of carbon dioxide to RuBP. You have now got a six-carbon molecule, which breaks into two identical three-carbon molecules. One of these molecules gets cycled back into RuBP and the other goes to make sugars for the plant. The plant then uses these sugars to make other molecules and to grow, producing leaves, fruits or seeds. We then harvest the plant and it gets eaten by you. The crucial and difficult step in all of this is the addition of carbon dioxide to the RuBP, and it can only happen because of a protein that helps the reaction along, an enzyme in other words, called ribulose bisphosphate carboxylase-oxygenase, which has the much handier acronym of RuBisCO.

Now we come to the big problem with photosynthesis: RuBisCO is a rubbish enzyme. Our bodies are chockful of enzymes – pretty much all of the proteins in our bodies or in plants are enzymes of one sort or another – and each enzyme is a specific helper for a specific chemical reaction. Most enzymes will churn through a thousand or so cycles of its dedicated chemical reaction every second. RuBisCO only manages a couple per second. Partly this is because it is dealing with a gas, carbon dioxide, but mostly it's just not very good at its job. If we could somehow tweak RuBisCO to make it more efficient then photosynthesis would run better, more sugar would be produced, the plant would grow faster and, bingo, you have bumper crop yields. Well, it has been tried and so far it only made RuBisCO even slower, but that has not stopped scientists from persevering in this area.

The most recent and best attempt to tackle this issue took a different approach. One of the reasons RuBisCO is slow is that it struggles to grab on to RuBP, the five-carbon molecule

THE FUTURE OF FOOD

that gets carbon dioxide stuck to it. By increasing the amount of the enzymes that make this precursor molecule, you make more RuBP and photosynthesis runs faster. Researchers in the UK genetically engineered wheat to do just this and, in the greenhouse at least, the wheat produced 20 per cent more grain. Which maybe doesn't sound much, but it is a huge advance. It means that where you used to have six fields of wheat, you would now only need five fields to make the same amount of grain. The crop is now undergoing field trials and, if successful, it will be the first crop specifically engineered to directly increase yield.

Unfortunately, there is yet another problem with RuBisCO, but it does have a silver lining. When RuBisCO evolved, some 3.5 billion years ago, the earth's atmosphere was chockful of carbon dioxide but contained no oxygen at all. This is not a problem for photosynthesis as it does not need oxygen. However, the process of photosynthesis does produce oxygen as a waste product. Over billions of years, all the tiny, single-celled creatures living on the products of photosynthesis began to raise the level of oxygen in the atmosphere. In fact, *all* the oxygen in our atmosphere is as a result of photosynthesis. And this is where photosynthesis hits a bump in the road. It turns out that evolution blindly created RuBisCO and not only was it good at adding carbon dioxide to RuBP, but it could do it with oxygen as well. It's actually in the name, ribulose bisphosphate carboxylase-oxygenase; the carboxylase bit refers to its ability to stick carbon dioxide to RuBP and the oxygenase to the fact that

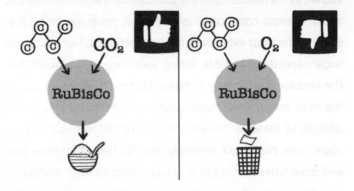

it adds oxygen. When oxygen is added to RuBP, the chemical pathways of photosynthesis are messed up. Rather than two identical three-carbon molecules, you get one three-carbon and one useless two-carbon molecule. All of which results in a big, big drop in efficiency.

That is not the end of it, though. The most exciting avenue for plant science right now is that evolution gave rise to a solution that plant scientists currently believe has arisen independently some sixty-one times. To keep your RuBisCO focused on carbon dioxide and ignoring oxygen, first isolate your promiscuous RuBisCO from the atmosphere by hiding it away inside cells surrounded by a protective ball of other cells. Your RuBisCO now has no oxygen to worry about, but also no carbon dioxide. Step two is to set up a transport system using a four-carbon molecule that ships the carbon dioxide through the protective ball of cells. The RuBisCO is bathed in lots of carbon dioxide and no oxygen. Plants that use this system are known as C4 plants, from the four-carbon transport molecule, and it is most common in grasses. It is no coincidence that some of the most extensively grown food crops, like maize and sugar cane, are C4 plants. It may seem like a lot of work, but the increase in efficiency of RuBisCO more than makes up for the extra energy spent transporting carbon dioxide around. It's difficult to measure the efficiency of photosynthesis, but C4 sugar cane plants are generally seen as being between two and three times as efficient as similar plants without the special modifications to their photosynthesis. Why then haven't C4 plants taken over and outcompeted regular plants? One side effect of all this extra shuffling carbon dioxide about is that you need more intense sunlight, although C4 plants are also better at coping with dry conditions. Plants that have evolved C4 photosynthesis tend to be found in hotter and drier climates where the sunshine is more intense.

It's C4 photosynthesis that is the silver lining to rubbish RuBisCO and where the latest plant science comes into play. It should be possible to genetically engineer C4 photosynthesis into plants that don't already have it. This is currently the holy grail of plant science as, if it can be done to something such as rice, the increase in yield from each field would be huge. The problem is that not only do we need to change the biochemistry of the plant, we need to change some basic structures as well. Before we can do that, we need to understand how these structures are created and coded for in the plant's genes, and that in itself is a massive challenge.

I may be somewhat biased, since I trained as a plant scientist many aeons ago, but I think this is one of the most exciting and significant problems facing science today. It is also probably the best chance we have of feeding the population of the world in the next fifty years.

When is green too green?

Messing about with the biochemistry insides plants may not be the only way to grow more food. Understanding how plants evolved helps, too. The process of natural selection gave rise to plants that are good at competing within their own little patch of the world for resources like space, water and sunlight. Part of being competitive is also the ability to grow quickly and that requires an efficient system of photosynthesis, but it is not the only way you can muscle out other plants and take over.

When plant scientists began to really pick apart how photosynthesis worked, they realized that many plants had too much of the green, sunlight-capturing pigment chlorophyll in their leaves. To recap, when sunlight hits a plant its energy is captured by the chlorophyll, which is then used as part of the process that feeds the RuBisCO enzyme to make sugars (see page 186). Specifically, a single photon of light is absorbed by a single molecule of chlorophyll, which in response spits out a single high-energy electron. That electron gets passed down a long chain of proteins and is eventually used to make a special energy-transport molecule. It is that transportable energy that drives the stuff that RuBisCO does.

The problem comes when too much sunlight falls on to a leaf. The chlorophyll starts spitting out lots of overly excited electrons and you end up with more electrons than the rest of the system can cope with. The surplus excited electrons then head off to wreak havoc elsewhere in the leaf. This is a common occurrence for plants as light levels are very variable, but the plant still needs to have an efficient photosynthesis machine that works on a cloudy day. So, on a sunny day, when the chlorophyll is in overdrive, a system called non-photochemical quenching comes into play. This mops up the extra electrons and turns them into heat. It's clearly a waste of the captured energy but it is better than the damage caused by an over-excited electron on the loose. It is also, and this is the kicker, a drain on the plant's energy as it needs to divert resources into the quenching system.

So, evolution gives rise to a clever system to protect the plant from excess light intensity. Huzzah, evolution saves the day! But evolution also gives rise to an overproduction of chlorophyll in the first place. If a plant can completely block out all the light with its leaves, then no useful light hits the ground beneath, and that means any smaller plants below it can't survive. The world of plant growth is just as vicious and cut-throat as the animal kingdom we see played out on TV documentaries about the plains of the Serengeti. If you can get an edge over other plants you can, quite literally, secure your place in the sun. Plants have evolved to produce more chlorophyll than they need, even on a cloudy day, because this way they can be nasty to smaller plants, which wither and die. Huzzah, evolution turns big plants into baby-plant-killing machines!

Back in the 1980s plant scientists were using a number of brute-force techniques to produce lots of random mutants of crop plants such as soybeans. Each plant created had a different mutation and was mostly either the same as the un-mutated plant or usually less healthy and a bit sickly. Occasionally, though, a plant would be made that was more vigorous in some way and stood out from the rest. Those plants were isolated and their genetics and biochemistry analyzed to work out what was different. In several cases, mutant soybeans were made that had less chlorophyll, but were counter-intuitively more prolific and gave higher crop yields. The thing is, soybeans grown in a field don't have to compete with a bunch of other weeds so they don't need to be baby-plant-killing machines. The use of modern agricultural techniques and the judicious application of specific herbicides leaves the plants free to spend all their energy on growing and making soybeans for us to eat. A mutant soybean with half the amount of chlorophyll produced nearly a third more crop than a plant with the normal level of chlorophyll. The mutant isn't producing nearly as many extra excited electrons and consequently doesn't need to expend all the energy on its quenching system. It may seem counterintuitive but this is probably going to be true for many of our common crop plants. If we can dial back the green we may end up with a bigger harvest and more to eat.

Cool new technology

The future of food is not going to be just about tinkering with plants and making pretend meat. The technology associated with our food is also going to be an exciting area to keep an eye on. Early in this book, I boldly claimed that one of the most important scientific and technological advances in food was the development of the refrigeration system and its use in our homes, shops and transport networks (*see* page 51). But what of the future for this revolutionary technology? We have, after all, been using essentially the same refrigeration system since 1805. There have been a few small tweaks to this system of using a pressurized gas to do the cooling; different motors have been fitted to our refrigerators, new and less polluting gasses used, but at its heart the science remains the same. That may be about to change if one or both of two completely novel technologies can make it out of the realms of specialist use and into mainstream cooling devices.

In 1881 at the University of Freiburg in the Black Forest of Germany, scientist Emil Warburg made a curious observation.

When certain metallic substances are placed in a magnetic field there is a short burst of very rapid warming. Similarly, when the metal is removed from the magnet, there is a little pulse of cooling. If, rather than using a permanent magnet, you use an electromagnet you can induce this cooling and warming effect by just turning the electromagnet on and off. But Warburg's observation of magnetic cooling, or the magnetocaloric effect, was mostly seen as a curiosity. For forty years it languished, unloved in the textbooks of physics until researchers working on extreme low temperatures worked out how they could use the effect to creep closer and closer to the ultimate cool of absolute zero (-273.15°C or -459.67°F). With the added interest of the experimental physicists, magnetic cooling became a viable, if niche, technology. Since then the principle has very slowly moved out of the laboratory and into industrial applications.

In 2016 French firm Cooltech, based in Strasbourg, released the first magnetic refrigeration system designed for use in shops and large industrial situations. Their system has a number of significant benefits over traditional compressed gas refrigeration. Firstly, and probably most importantly, is that it uses less energy to produce an equivalent amount of cooling. On top of that, it is less noisy and since it contains no pressurized gas there is no risk of accidental leakage. Which is a good thing as, even though we no longer use the refrigerant Freon that was destroying the ozone layer, the current refrigerant gas most common in the UK, called isobutane, is still a greenhouse gas about three times as harmful as carbon dioxide. Cooltech's system relies on water-based circulation in pipes to cool the inside of the refrigerator. The actual cooling is done using a cunning double-rotating system of magnets. Crucial to the technology is the use of the rare elemental metal gadolinium, which gives a particularly strong magnetic cooling effect. The spinning magnets repeatedly cool the gadolinium, which is bathed in a flow of water. In turn, the gadolinium cools the water that is then pumped away to keep the refrigerator cold. The company claims that their magnetic system uses only half the energy that a traditional system uses. When you consider that globally about a fifth of all the energy we create goes into cooling systems, then this could be a very significant saving. This is especially true when you add in predictions that the demand for cooling, mostly in the form of air conditioning, is set to massively increase in the next few decades. In the European Union, by the year 2030 the total bill for energy used for cooling is set to increase by over 70 per cent. Clearly this is a promising bit of new tech and naturally at this stage it is still very expensive, but given time the cost will come down, at which point it will become viable for domestic use.

If you don't fancy using magnets to keep you or your food cool, then a company based in Wales called Sure Chill have an equally clever but lower-tech solution. I had a chance to meet and work with the inventor, Ian Tansley, while filming for a television programme. I had previously met Ian a long time before while working on another programme about environmentally sound solutions to everyday living. Then he had driven over to the filming location with his recycled chip-shop-oil-fuelled car. It had a distinct aroma that left me craving fish and chips all day. It transpired that after our first encounter, some time in 2008, he began thinking about refrigeration when walking in the Welsh mountains with some chums. On passing a frozen lake, he started trying to explain to his friends why only the surface of a lake freezes. This was his eureka moment, so the story goes, but I suspect there was a lot of hard graft involved, too.

The key to Ian's invention is the peculiar nature of water. In fact, it has such bizarre properties that it is a very bad example of how a liquid should behave, which is ironic as it is by far and away the most common liquid we encounter in our everyday life. Consequently, it does not occur to us that solid ice floating on top of liquid water is freaky behaviour. Normal, well-behaved substances get more and more dense as they cool. Or to put it the other way around, they shrink as they get colder and the same amount of stuff is packed into a smaller space. So, for example, imagine all the atoms in a cupful of liquid mercury: as the mercury cools the atoms jostle about less, pack together closer and the mercury takes up less space in the cup. When mercury freezes into a solid at -39°C (-38°F), the atoms pack even tighter and there is an even more drastic reduction in volume which translates to a big jump up in density. Solid mercury does not float on top of liquid mercury, it sinks to the bottom. And this is what most well-behaved substances do. In fact, not doing this is super rare and only seen in a handful of substances. Water just happens to be one of them.

As water cools it does indeed slowly become more and more dense, until it hits 4°C (39.2°F), at which point it gets weird. Water is capable of making what are known as hydrogen bonds; in fact, it's really, really good at making them. To make a hydrogen bond you need two things: a slightly positively charged hydrogen atom and a different atom, usually oxygen, with a slightly negative charge. You can get these two things if you take a hydrogen atom that is connected to an oxygen atom. The oxygen pulls the electron in the hydrogen slightly away from the hydrogen. This leaves the hydrogen slightly positively charged and the oxygen slightly negatively charged. Since water is made of just two hydrogen atoms stuck to a single oxygen atom, it is ideally placed to make hydrogen bonds. Except that hydrogen bonds are weak. In warm water the molecules are zooming about with too much energy to be bothered with feeble hydrogen bonds. Until, that is, you get to the magic temperature of 4°C (39.2°F). At this temperature, the energy within the water molecules is low enough that hydrogen bonds start to form. When this happens, the water begins to arrange itself into a more ordered but still fluid structure, and in so doing the molecules space themselves out a bit. If your molecules are spaced out and packed less tightly, your density starts to drop. As water gets colder than 4°C, it becomes less dense and expands. When it freezes into ice, the hydrogen bonds take over and the density of the water,

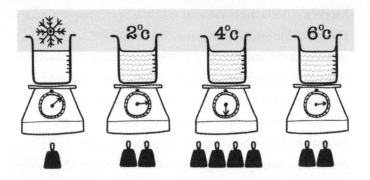

now ice, drops dramatically. When water freezes it expands by nearly 10 per cent. Which explains why pipes burst in winter and why ice is so freaky and floats on water.

How, then, does this help make a refrigerator? When Ian Tansley read around the subject a bit more he realized that the science going on was perfect for refrigeration. The temperature of the water in his frozen lake in the Welsh mountains, or any lake for that matter, is always at 4°C (39.2°F), aside from the water immediately next to the ice. You don't get a temperature gradient in the water in a lake; it's all at 4°C from top to bottom. So long as some of it is liquid, the water will be at a constant 4°C because this is the temperature at which the water is at its maximum density. If something warms up a portion of the water in the lake, say a fish does a bit of swimming and creates some heat, the warmer water gets less dense. If the water gets less dense it begins to float up to the surface of the lake, hits the ice and is cooled. When it gets to 4°C it is now denser than the cold ice at the top so it sinks back down. Similarly, imagine what happens if something were to cool the water halfway to the bottom of the lake. As the water cools below 4°C, once again it gets less dense and floats up to the surface where it stays. The bulk of the water stays at the magic 4°C, no matter what you throw at it, so long as there is a layer of solid ice on top.

It is this peculiar behaviour of water that led to the creation of Sure Chill's first prototype refrigerator. At the top, you have a large compartment full of water and the cooling coils of a conventional refrigerator. Below this is a series of pipes that allows water to freely move around the body of an insulated, open space into which you can put items to be chilled. It looks basically like a regular refrigerator, with a door and shelves, but there is a bit of space at the top you can't use as that's where the water and cooling coils are. You start the refrigerator by turning on the conventional cooling system and freezing the water in this top section solid, to form a big block of ice. Anything in the lower storage compartment now becomes chilled as the surrounded pipes fill with water at 4°C (39.2°F), which just happens to be the perfect refrigerator temperature. This is where it gets really clever; you can now turn the refrigerator off and disconnect it from all power sources. So long as you properly insulate the big block of ice at the top, the refrigerators made by Sure Chill will stay cold for up to ten days. Now, admittedly, what they don't say is that this assumes you only open the refrigerator a couple of times and you don't put lots of warm groceries in, but even if you did it would still go for a couple days without power. The block of ice will very, very slowly melt, but all the while the refrigerator stays at 4°C. You don't even need any power to monitor the temperature inside to make sure it stays that way. Due to the quirks of water's maximum density that is the only temperature the inside of the refrigerator can possibly be at.

Ian Tansley originally designed this refrigeration system not for domestic use, but for vaccine storage in developing countries. Vaccines that are needed for eliminating terrible diseases in places such as Africa are very heat sensitive. They need to be kept refrigerated at all times and even the most efficient refrigerators need to have a constant supply of electricity. In many of these developing countries, power cuts are frequent due to a chronic shortage of electricity, so a refrigerator that can cope for long periods with no electrical power is life-changing.

In 2014 the Bill and Melinda Gates Foundation got in touch with Sure Chill and set them a challenge: to invent a refrigeration system that could store vaccines for a month without a power source. The Foundation also gave them a healthy pot of cash to help with it. Three years later the challenge was complete. Right now, as you read this, there are refrigerator systems in thirty-eight different countries, mostly in Africa, that are designed specifically to store a small box full of vaccines at a constant 4°C (39.2°F) for a whole month, without any input of electricity or power. The refrigerators work in temperatures up to a sweltering 43°C (109°F). The ice inside comes in packs that can be frozen elsewhere, transported to the device and popped in the back. So, you don't even need a source of electricity where the refrigerator is located, so long as, once a month, somebody tops up the ice packs.

So, when can you buy one of these refrigerators? I hear you ask. Well, it is going to be some years yet. For speciality, niche uses both magnetic and ice refrigerators are already out there in the wild being used. But the problem is, as with any new technology, you don't just have to convince the consumers that it is the way forward; manufacturers of conventional refrigerators will need to be sure that they can make a profit from day one. They are not going to commit to turning over a whole production line to a new technology if they are not certain it will be a success from the start. I figure it will be five years before you can buy a domestic version of either Cooltech's magnetic or Sure Chill's ice refrigeration system, and that will only be high end, speciality products for the tech geeks who are willing pay a premium for the boost in energy efficiency. For the rest of us, give it ten years and the technology inside your refrigerator will have completely changed.

Future farming

Here's the bottom line: by the year 2050 the population of the planet will be just short of 10 billion people and all of these people will need feeding. According to the United Nation's Food and Agriculture Organization, this translates into a 70 per cent increase in agricultural production from 2009 levels. The problem is that almost all of the land on the planet suitable for agriculture is already being used to grow food. While it may be possible to boost the productivity of some of this land by applying the latest best-practice ideas for traditional farming, for many crops like rice and wheat we are already doing this and growing as much as we can. If we want to grow enough food to feed the planet, we need to start thinking differently about agriculture.

One area of advance that is already proving to be both successful and profitable is what has been dubbed smart farming. Traditionally, a farmer would use their experience to gauge the best moment to water, fertilize or apply pesticides to a field full of crops. In so doing they would weigh up lots of variables to get the best yield from their resources. What smart farming can do is provide much more information on a smaller scale. Take almonds, for example; these are a notoriously water-hungry and tricky crop to grow successfully. Farmers in California are now fitting moisture sensors in the soil next to every tree in their groves. The data from these sensors is then wirelessly sent to a central computer system that works out how much water each tree needs for optimum crop yield. A tree-specific water dose is then delivered direct to each tree through automated pumps and a system of hoses. The result is a bumper yield of almonds and a one-fifth saving in water.

New sensors being developed now will soon allow the almond farmer to detect other nutritional requirements of their trees like available nitrogen in the soil, which in turn can be used to deliver customized doses of fertilizer.

Fitting sensors into the ground around long-lived trees makes perfect sense – the investment pays itself off over the years of use – but how do you collect equivalent data from a field of millions of wheat plants? One solution being piloted by the University of Nebraska in the United States is to use the machines already employed by the farmer. Modern tractors are routinely fitted with global positioning systems (GPS) that allow the drivers to see, for example, which bits of the field they have sprayed and which bits remain to be done. By fitting soil sensors to the tractor, it is possible to gather information about moisture that can then be linked to the GPS data to create a map showing which plants need watering and which don't. The water can then be delivered by computer-controlled, GPS-carrying, robotic watering vehicles to exactly where it is needed.

Once we start moving into the era of autonomous robots we open up the possibility of using drones with high-resolution cameras to take an aerial view of a whole field. Such systems are already being developed in Australia at the University of Sydney, where drones are used to locate weeds. Once the weeds have been identified from aerial images, the second part of the system comes into action. The drone passes on this information to a four-wheeled, solar-powered robot called RIPPA (Robot for Intelligent Perception and Precision Application). RIPPA is designed to be able to drive along rows of vegetables using the tractor tracks already in the field, so there is no disturbance of the crop. Once it reaches an area of interest, cameras and intelligent learning systems identify the weed, and a controlled and directed squirt of herbicide is applied. The latest development of RIPPA includes a moisture probe so the robot can gather data for tailor-made watering as well.

All of this technology allows the farmer to increase yields by giving their plants the optimal growing conditions. It's almost as if we are creating a bespoke farming system for each and every plant. Clearly, at this stage such systems are expensive and early use is limited to high-value crops like almonds. But as the technology is developed, costs will come down and eventually be outstripped by increased resource savings and higher yield values.

If farming in a field has the potential to be revolutionized then so does growing in a greenhouse. Since within a greenhouse you can control more of the growth environment, it opens up the possibility of growing crops in inhospitable areas. An enterprise called Sundrop is doing just this in Port Augusta in South Australia. This area of Australia is incredibly arid and technically a desert with maximum summer temperatures just below 50°C (122°F). Unsurprisingly, there is not much agriculture in the area, so you certainly would not expect it to be a location for a huge 20-hectare (about 50-acre) greenhouse full of tomatoes. The secret to Sundrop's success is an equally large solar-collection system. A huge field of mirrors focuses the ample Australian sunlight onto the top of a tall tower. The collected solar heat is used to boil seawater, creating steam that powers turbines and makes electricity. This solar electricity is then used to desalinate seawater, pump the freshwater into the greenhouses and, when needed, run air-conditioning units to keep the temperature just right for the tomato plants. As an added twist, the plants are grown using the principles of hydroponics where the roots are just bathed in water with a few nutrients dissolved in it. No soil is needed. Taken together, the Sundrop system is a complete self-sustaining package that can be set up in areas previously seen as unsuitable for crops like fruit and vegetables. The enterprise is in the process of expanding into Portugal and Tennessee in the United States.

The hydroponic system used by Sundrop is not a new way to grow plants. It was first experimented with by the great British philosopher, scientist and statesman, Francis Bacon, in 1627. What it boils down to is that plants do not need soil to survive. The roots of a plant provide a couple of vital functions. Firstly, they are needed to draw water and a relatively sparse range of nutrients into the plant. You don't need soil to do this, just lots of water. The other function of the roots is to anchor the plant and hold it upright so that its leaves are presented to the sunlight. The roots have evolved to do this by growing into and then holding onto the soil. But if you remove the necessity for this second function, by creating artificial supports for the plant stem, you can get rid of the soil. The now bare roots can be bathed in water containing just the right levels of minerals.

Hydroponics has developed a long way since Francis Bacon and is now routinely used in greenhouses for the growth of much of our salad vegetables like peppers and lettuces. It has also spawned a less well-known but potentially interesting system called aeroponics.

While in hydroponics you grow plants with the leaves in the air and the roots submerged in a water bath, aeroponics does away with the water bath. Instead, the roots are held in a sealed chamber filled with just a mist of water droplets that contain nutrients. While it is slightly more technically challenging to set up, it means you need a fraction of the water used for a hydroponic system. On top of this, roots tend to grow a bit better in aeroponic systems as there is an increased availability of oxygen. From a farmer's perspective, another advantage is the reduced weight of an aeroponic growth system and that allows for vertical farming.

Several companies around the world are now producing and selling vertically farmed crops. At the moment, it's only lettuce that is being grown, but there are ambitious plans for other food plants to be added. Vertical farming adds two new elements to the modern farm. The crop is sown onto a thin cloth stretched across the top of a large deep tray. The space between the cloth and the bottom of the tray is the aeroponic chamber and is filled with a nutrient mist. Once the seed germinates, the roots grow through the cloth and into the aeroponic chamber. Since the trays are just full of air, they can be stacked one on top of each other, leaving a gap for the plants to grow in. The problem now is how to get light to the growing seedlings and that's where the latest optical-electronic innovations help out.

The light emitting diode (LED) has become a commonplace electrical component found in everything from head torches to television displays. The latest developments have enabled scientists to create LEDs that emit very specific colours of light, which is important if you want to grow lettuce. Plants are mostly green due to the pigment chlorophyll in their leaves (see page 192). If something looks green, it must be reflecting green light and by the same account it must also be absorbing all the other colours of the spectrum, like red and blue light. This is exactly what green plants do. So, when it comes to photosynthesis (see page 192), green light is irrelevant and it is specific wavelengths of blue and red that plants want. Vertical farms illuminate their aeroponic trays by adding red and blue LEDs to the undersides of the trays above. You now have a system that provides water, nutrients and light to the plants, and is sufficiently lightweight that it can easily be stacked into huge banks of trays. If you have a stack of two aeroponic trays on top of each other, you have just doubled the growing capacity of the bit of land the trays sit on. That may not seem that impressive, but there is no need to stop there. Ten, twenty or more trays could be vertically farmed and, since you don't need sunlight, you can set up your farm anywhere. For example, there is a vertical farm called AeroFarm currently working in an old warehouse in downtown Newark, just a short way from central New York City in the United States. It produces high-quality, baby salad greens for the local community. On the outside, there is nothing about the warehouse to distinguish it from all the other buildings in the area, and yet inside, bathed in purple light (a mix of blue and red), is the future of farming. If we want to feed the future population of the world we will need radical ideas like vertical or smart farming. Crucial to this process will be an understanding of how plants grow and what they need to grow, and then giving them each exactly what they want so that we can harvest the biggest crops from the smallest spaces.

Fighting food fraud

There is one final aspect of the future of food it would be remiss of me not to mention. Technology is allowing us to do ever more incredible things with the food products we eat, but can we trust that these products are what they say they are? If you are going to eat a beef burger, or even an artificially grown burger, can you be sure that the meat comes from the stated source?

One of the most publicized examples of food fraud started on 15 February 2013 when the Food Safety Authority of Ireland announced that horsemeat had been found in frozen beef burgers that were on sale in a number of major supermarket

THE FUTURE OF FOOD

chains. The situation rapidly developed as similar contamination was confirmed in the UK and then, over the course of just a month, in thirteen more countries in the European Union. Apart from the yuck factor of having unwittingly ingested horsemeat, there were also concerns about the quality of that horsemeat. There are common veterinary drugs used on horses that are prohibited within the human food chain and any horse treated with these drugs cannot be used for human consumption in any way. It transpired that in the 2013 case some very low-level contamination by a horse drug called phenylbutazone had occurred. The circumstances were further complicated when it came to light that many other processed beef products were also contaminated with pork meat. Depending on your cultural and religious background, this may also raise serious concerns; for example, both horsemeat and pork are taboo meats for Muslims and Jews. But there is an underlying principle at stake here for us all; if it says beef on the packaging, then we should be able to trust that this is what we will be consuming. The horsemeat scandal also raised other troubling questions. If horsemeat was found in what we supposed was a wholly beef product and not included in the list of ingredients, what else might there be in our food products that manufacturers are failing to mention?

This is where exciting new technologies can come to our rescue. The key to detecting food fraud is speed. Since many food products have a limited shelf life, if you are going to develop a testing method it needs to be quick. The window of opportunity may only be a few weeks or even days before the food has been bought, eaten and the potential damage done. Testing needs to give immediate results for it to be useful. It also needs to be a generic test capable of identifying a whole range of contaminants since you don't know what you are looking for when you perform the test.

In 2012, a Hungarian chemist called Zoltan Takats realized that a medic's electrosurgical knife could be used as a diagnostic tool as well as an implement for cutting. This type of knife has been around since the 1920s. It relies on an electric current to cauterize the tissue as it is being cut. What Zoltan realized is that the smoke produced by electrosurgical knives was a source of biological information. In particular, the smoke contains vaporized fatty acids from the membranes of the cut cells. A small tube placed next to the tip of the electrosurgical knife can suck up all those vapours and direct them into a fancy bit of laboratory kit called a mass spectrometer. Taken together, the knife and mass spectrometer have been dubbed the iKnife and they allow you to work out the relative proportions of a complex mix of vaporized molecules. Zoltan's interest was in using this smoke to identify cancerous cells that show a slightly different profile of fatty acids to healthy cells. However, it soon became clear that every species of plant or animal gives a slightly different profile of fatty acids, and therefore the iKnife could be used to distinguish the presence of different foods. It's an ideal test for fraudulent food as the results take but a few moments to gather. It's a new technique so it is only just coming into general use, but a 2015 study on fish sold in the Boston area of the US found that nearly half of all food shops tested were selling mislabelled fish of some sort. Of the red snapper fish tested, only one in twenty was actually red snapper, the rest being a related but less desirable species. More worryingly, it was found that about half of the tuna sold was not tuna but came from a fish called escolar. This unrelated species contains a fatty acid that, after even quite small portions are eaten, can cause stomach cramps and diarrhoea. Clearly, mislabelling can often spell bad news for the consumer.

Other things that lend themselves to analysis under the iKnife have included butter from 'grass-fed' cows that in reality never saw any grass, and identifying manuka honey that wasn't from manuka trees. Anything wet and fairly simple can be analyzed this way. For more complex mixtures, though, a fatty acid profile alone can be insufficient to pick apart the origins of everything in the mix. Which is why food-fraud experts often turn to DNA analysis. The DNA code contained within any cell isn't just slightly different to cells from other species, it's completely different. To analyze the DNA, you need to use a technique called DNA sequencing that allows you to work out all the letters in the DNA code. When this was first developed it was a laborious and slow process. One of its greatest achievements was the Human Genome Project that set out in 1990 to completely map the genetic code of a human being. It was an international project involving thousands of scientists, dozens of scientific institutions and took thirteen years to complete. With today's technology, you can do the same thing with one person in a matter of hours. It is quite a staggering increase in the capability of the technology. These days, the issue is not so much about if you can get a DNA sequence, but what you do with the vast mounds of data you produce. Thankfully, developments in computer science and mathematical algorithms are helping biologists make sense of it all.

The most exciting new DNA-sequencing gadget is made by a UK-based company called Oxford Nanopore Technologies and is about the size of a large USB memory stick that plugs straight into your computer. You load the sample directly into a tiny little chamber inside the machine and within a few hours you have a DNA sequence. It is ideal for analyzing suspect food samples. Since it is completely portable it can be taken wherever it is needed, rather than having to send samples to a laboratory. What is more, since it does not rely on wet samples and the fatty acids within them, you can test dry powders.

One of the more common types of food fraud committed on the public is fake spices. Some, like adulterating peppercorns with papaya seeds, are easy enough to detect. In this case, peppercorns sink while papaya seeds float. But it can be more difficult with something like saffron powder, which is often adulterated with turmeric. Saffron comes from the stamens of a particular variety of crocus flower, whereas turmeric is from a completely different and much cheaper to grow plant root. Saffron is also the most expensive spice that you can buy, which only adds to the temptation to adulterate it. The addition of just one in ten parts turmeric can massively boost profit margins. But a bit of DNA sequencing can tell you if your saffron is mixed with turmeric root.

The same methods can be used to police the trade in internationally banned food products. Most orchid species are protected under the Convention on International Trade in Endangered Species (CITES), but this has not stopped a huge market in the underground storage body, or tuber, of orchids. Flour made from these tubers is used in the production of a number of Turkish and East African delicacies. A particular variant of starch in the tuber flour gives a glutinous texture that makes these delicacies unique (*see* page 61 for more starchy variants). Sadly, the orchids used to make this flour are not being farmed, but taken directly from the wild. The harvest is now a major source of income for some of the poorest people on the planet. Unfortunately, several East African orchid species are now on the brink of extinction. Using rapid DNA sequencing, samples of flour can be examined and the exact species contained determined and also where the orchids originated. Which in turn can help botanical policing and also direct the efforts of education to the communities damaging the wild populations of plants. With the help of rapid DNA sequencing, the hope is that the illegal trade can be slowed and new farming projects set up in the communities that need the income from orchid tubers to survive.

I have spent quite a while in this book telling the stories of how our food and food technology got to where they are today. I've always believed that the history of science allows us to tell stories that not only bring the subject to life, but help explain the current state of our understanding. However, I wanted to end with a look to the future, which has its hazards. Only time and hindsight will tell which of the current crop of new and exciting ideas, some of which I have talked about in the previous chapters, will turn out to be the food pills of the year 2050. The latest technologies and developments in both plant and food science are allowing us to open up the varieties and range of foods that we can choose from. While inevitably this brave new world of opportunity is abused by individuals who seek to pass off fraudulent food, we also now have the technology to track and monitor what goes into the food we buy. Looming on the horizon there are some serious and major food problems we need to deal with on a global scale, but the ideas and technology in development right now have the potential to tackle many of these issues.

I think the twenty-first century is going to continue to be an exciting time for the science of food.

Acknowledgements

While the act of writing a book may seem like a solo activity, and maybe for some it is, for me it is one of collaboration. To that end, I need to thank the people who helped me bring this tome to life. My agent Sara Cameron and her colleagues Vicki McIvor and Ruth Smith are always there to help out, and Sara in particular copes with all the panic and stress I throw at her when I'm running over schedule.

Talking of being over schedule, the folks at Michael O'Mara have been extraordinarily patient with me once again, for which I am very grateful. While this time around I had some clue as to what needed doing, other work and some procrastination stretched out the time taken to write the book. To be honest, there was quite a lot of procrastination.

The greatest thanks go to my family. My children had to endure me loitering in the house when I was writing and I apologize for hassling them every time they came home from school. My wife, Juliet, was vital to the process, constantly supplying me with interesting nuggets of food science. Most importantly, she is also responsible for making sure I'm not talking complete nonsense.

Index

(page numbers in italics refer to illustrations)

A

adenosine 129–31
AeroFarm 209–10
aeroponics 208–10
alginate 67–9
allergies 180–6
Amadori compound 102
amino acids 28–9, 102–3, 120,
    123, 172–3
amphiphiles 88–9, 107
Appert, Nicolas 144–6
Arrhenius equation 43, 51–2, 82
aspartame 97–8

B

Bacon, Francis 207
bacteria (*see also* hygiene;
    microbiota):
    Campylobacter jejuni 134
    on chopping boards 21–3
    Escherichia coli 134–5, 137
    and fungi, *see main entry*
    in gut, and change of diet 142
    in and on human body 137–42,
        165
    importance of 137
    killing 143–9
    Lactobacillus 157
    in mice 140–2
    microscope enables sight of
        143–4

and pasteurization 146–9, *147*
and refrigeration 52
salmonella 21, 134, 152
best-before, sell-by, use-by dates
    150–4
Bill & Melinda Gates Foundation
    202
Birdseye, Clarence 54
bread 70–7
    and Chorleywood process 72,
        73–7 *passim*
    and fermentation 158
    and gliadin 71
    and gluten 71–3
breakfast cereals 58–61, *59*
British Baking Industries Research
    Association 73
Brussels sprouts 121–6

C

caffeine 126–32 (*see also* coffee)
    and adenosine 129–31
    effects of 126, 130–2
    in tea 127
Campylobacter jejuni 134
canned food 145–6
caramelization 109–13
carrageenan 65–6, *65*
Catts, Oron 167
ceramic knives 24–5 (*see also*
    knives)

cheese, processed 90–3
chocolate 113–21, *119*, 155
chopping boards 19–24
    and bacteria risk 21–3
    and hygiene 21–3
    and Mohs scale of hardness
        20–1, *20*
    scientific studies into 21–3
Chorleywood process 72, 73–7
        passim
Churchill, Winston 166
Clarke, Jillian 135, 136
cocoa 114–18, *119*
coffee 81–2, 83–4, 126–32 (*see
        also* caffeine)
    beans 127–9
cooking temperatures 27–39, *28*
    and fats 27
    and properties of heat 30–5
    and protein 27–9, 35–6
    and sous vide method 35–9, *36*
Cooltech 196–7, 203
crystalline complexity 113–21
Cullen, William 54
cutting and shearing 14–19, *19*
cyclamate 96

date-sensitive food 150–4
Delightes for Ladies (Platt) 45
drum dryers 79–80, *80*
Durand, Peter 145

eggs:
    beating, *see* whisks and
        whisking
    and emulsifying 86

    and sous vide method 37–8
emulsifying 86–92, *86*
Escherichia coli 134–5, 137
Evans, Oliver 55

F

Fahlberg, Constantin 94–5
farming:
    and aeroponics 208–10
    future of 203–10
    and hydroponics 207–8
fats 106–7
    and flavour 105–9
    and oils, difference between 106
    and temperature 27
fermentation 70, 154–8
    and fungi 154–8
    and lactic acid 156–7
    and synthetic meat, *see main
        entry*
    and yeast 71, 72, 74, 157–61,
        161
fibre 76, 140, 146, 165
flavour 100–26
    and Brussels sprouts 121–6
    and caramelization 109–13
    and chocolate 113–21
    and cocoa 114–18
    and crystalline complexity
        113–21
    and fats 105–9
    and genetics 122–3, 124
    and Maillard reaction 39, 44,
        100, 102–5, 108
    and smell 101–2, 107–8
    types of 100–1
    and umami 100–1
food fraud 210–15

food pills 163–6
  in science fiction 164
food preservation 51, 53–4,
    143–9 (*see also* fermentation;
    refrigeration)
  canning 145–6
  pasteurization 146
food rheology 63
Frankenfood 170 (*see also* synthetic
    meat)
Franklin, Benjamin 54–5
freeze-drying 83–4
frozen-food industry, *see*
    refrigeration
fructose 102, 110–11, 112 (*see*
    also sugar; sweetness)
fungi 154–61 (*see also* fermen-
    tation; Marmite; Quorn)
  yeast 157

G
ghrelin 141
Girard, Philippe du 145
gliadin 71
glucose 61–2, 63, 64, 102, 103,
    110, 112, 155 (*see also* sugar;
    sweetness)
glucosinolates 123
gluten 71–3
  and Chorleywood process 74
gun puffing *59*, 60 (*see also*
    breakfast cereals)
Gutbrod, Georg 40–1

H
Harrison, James 55–6
Hodge, John 102
Hooke, Robert 39

Human Genome Project 213
hydroponics 207–8
hygiene 133–7 (*see also* bacteria;
    refrigeration)
  and chopping boards 21–3
  and dropping food on floor
    133–7
  and fungi, *see main entry*
  and killing bacteria 143–9
  and pasteurization 146–8, *147*

I
insects, eating 176–8
instant food 78–85 (*see also*
    processed food)
  mashed potato 78–9, *80* (*see*
    also drum dryers)
Inuit 54

K
kitchen equipment:
  chopping boards 19–24
  knives 14–19, *19*, 24–6
  pressure cookers 39–44, *41*,
    82
  and refrigeration 51–7 (*see also*
    main entry)
  and sous vide method 35–9
  and temperature 27–35 (*see also*
    cooking temperatures)
  and whisking 44–51, *45*
knives 14–19, *19*
  ceramic 24–6
  Santoku 14, 18–19

L
lactic acid 156–7
Lease, Mary 164

lecithin 86
Liebig, Justus von 159
Lindt, Rodolphe 114–15
Lysenko, Trofim 73

 M

Maillard reaction 39, 44, 100,
  102–5, 108, 111, 128
maltodextrin 64–5 (*see also* starch
  and thickeners)
Marmite 159–60
mashed potato 78–9, *80* (*see also*
  drum dryers; instant food)
Mattes, Richard 107
Meat Research Institute 108
meat substitutes, *see* synthetic
  meat
Memphis Meats 169–70
methylcellulose 67
micro-organisms, eating 159–62
  (*see also* bacteria)
microbiota, *see* bacteria
milk:
  and lactic acid 157
  and pasteurization 146–9, *147*
    (*see also* bacteria; hygiene)
  powdered 80–1 (*see also*
    processed food)
Miranda, Robyn 133–5, 137
Mitchell, Edward Page 163
Mohs scale of hardness 20–1, 25
Mottram, Don 108
myosin 29–30

 N

nicotinamide adenine dinucleotide
  (NAD) 156

 O

oil, non-polar molecules within
  87
oil–water mix 85–8 (*see also*
  emulsifying)
  and amphiphile 88–9
oleogustus 107
ovalbumin 28
Oxford Nanopore Technologies
  214

P

pans:
  and heat 30–5
  Teflon 33–4
Papin, Denis 39
Pasteur, Louis 146
pasteurization 146–9, *147* (*see also*
  bacteria; food preservation;
  hygiene)
peanuts, allergic reactions to
  180–1, 182–3, 184–6
phenethylamine 120–1
phospholipids 106–7
photosynthesis 186–95, *190*, 209
Platt, Hugh 45
Plunkett, Roy 33–4
polytetrafluoroethylene (PTFE), *see*
  Teflon
Post, Mark 167, 168, 169
powdered milk 80–1 (*see also*
  processed food)
pressure cooking 39–44, *41*, 82
processed food 58–99 (*see also*
  instant food)
  bread 70–7
  breakfast cereals 58–61, *59*
  cheese 90–3

coffee 81–2, 83–4
  and drum dryers 79–80, *80*
  and emulsifying 86–92
  and freeze-drying 83–4
  and gliadin 71
  and gluten 71–3
  instant 78–85
  and spray dryers 80–2
  and starch and thickeners
      61–70, *65*
  and sweetness 93–9
protein (*see also* fungi; Quorn):
  and amino acids 28–9
  amino acids in 28–9, 102, 120,
      123, 172–3
  daily consumption of,
      recommended 172
  enough to feed the world
      172–80, *179*
  glutenin and gliadin 71
  insects as 176–8
  myosin 29–30
  new sources of 161 (*see also*
      food pills)
  ovalbumin 28
  plant sources of 173–4 (*see also*
      photosynthesis)
  and sous vide method 35–9, *36*
  and temperature 27–9, 35–6
PTFE, *see* Teflon

Q
Quorn 160–2, *161*, 170, 178

R
refrigeration 51–7 (*see also* food
      preservation)
  and Arrhenius equation 51–2
  and bacteria 52
  future of 195–203, *200* (*see also*
      Sure Chill)
  and Inuit 54
  and vaccine storage 202–3
  yakhchals 53–4, *53*
Remsen, Ira 94–5
rheology 63
RuBisCO 188–93, *190* (*see also*
      photosynthesis)
Running, Cordelia 107

S
saccharin 94–6
salmonella 21, 134, 152
Santoku knife 14, 18–19
Schaffner, Donald 133–5, 137
Schlatter, James 97
sell-by, use-by, best-before dates
      150–4
Shiga toxin 134
short-chain fatty acids (SCFAs)
      139–41
sous vide method 35–9, *36*, 103
Spallanzani, Lazzaro 143–4
spirulina 178–9, 180
spray dryers 80–2
starch and thickeners 61–70, *65*
  alginate 67–9
  carrageenan 65–6, *65*
  maltodextrin 64–5
  methylcellulose 67
sucrose 102, 110, 112 (*see also*
      sugar; sweetness)
sugar 102–3 (*see also* sweetness)
  and caramelization 109–10
  and cocoa 114–15
  fructose 102, 110–11, 112

glucose 61–2, 63, 64, 102, 103, 110, 112, 155
  molecules 102
  substitutes for 94–9
  sucrose 102, 110, 112
Sundrop 206–7
Sure Chill 197–8, 201–3
Sveda, Michael 96
sweetness 93–9 (see also sugar)
  artificial, and health issues 97
  aspartame 97–8
  and bodyweight 98
  in cola 96
  cyclamate 96
  saccharin 94–6
synthetic meat 160–2, 166–71, *169*
  (see also protein; Quorn)

T

Takats, Zoltan 212
Tansley, Ian 197–8, 200, 202
taste, see flavour
Teflon 33–4
temperatures 27–39, *28*
  and fats 27
  and properties of heat 30–5
  and protein 27–9, 35–6
  and sous vide method 35–9, *36*
theobromine 118–20 (see also chocolate; cocoa)

thickeners, see starch and thickeners
tinned food 145–6

U

umami 100–1
use-by, best-before, sell-by by dates 150–4

V

vacuum cooking 35–6
Vischer, Alfred 41

W

Warburg, Emil 195–6
water, polar charges within 87
water–oil mix 85–8 (see also emulsifying)
  and amphiphile 88–9
whisks and whisking 44–51, *45*
  and choice of bowl 49

Y

yakhchals 53–4, *53* (see also refrigeration)
yeast 71, 72, 74, 157–61, *161*
yeast extract, see Marmite

Z

zirconia, knives made from 24–6
Zurr, Ionat 167